ENGINEERING STRESS ANALYSIS:
A Finite Element Approach
with FORTRAN 77 Software

ELLIS HORWOOD SERIES IN MECHANICAL ENGINEERING

Series Editor: J.M. ALEXANDER, Professor of Mechanical Engineering, University of Surrey, and Stocker Visiting Professor of Engineering and Technology, Ohio University

The series has two objectives: of satisfying the requirements of post-graduate and mid-career engineers, and of providing clear and modern texts for more basic undergraduate topics. It is also the intention to include English translations of outstanding texts from other languages, introducing works of international merit. Ideas for enlarging the series are always welcomed.

ENGINEERING STRESS ANALYSIS:
A Finite Element Approach
with FORTRAN 77 Software

D. N. FENNER, B.Sc.(Eng.), A.C.G.I., Ph.D., D.I.C., M.I.Mech.E., C.Eng.
Department of Mechanical Engineering
King's College London (KQC), University of London

ELLIS HORWOOD LIMITED
Publishers · Chichester

Halsted Press: a division of
JOHN WILEY & SONS
New York · Chichester · Brisbane · Toronto

First published in 1987 by
ELLIS HORWOOD LIMITED
Market Cross House, Cooper Street,
Chichester, West Sussex, PO19 1EB, England
The publisher's colophon is reproduced from James Gillison's drawing of the ancient Market Cross, Chichester.

Distributors:

Australia and New Zealand:
JACARANDA WILEY LIMITED
GPO Box 859, Brisbane, Queensland 4001, Australia

Canada:
JOHN WILEY & SONS CANADA LIMITED
22 Worcester Road, Rexdale, Ontario, Canada

Europe and Africa:
JOHN WILEY & SONS LIMITED
Baffins Lane, Chichester, West Sussex, England

North and South America and the rest of the world:
Halsted Press: a division of
JOHN WILEY & SONS
605 Third Avenue, New York, NY 10158, USA

© 1987 D. N. Fenner/Ellis Horwood Limited

British Library Cataloguing in Publication Data
Fenner, D. N.
Engineering stress analysis: a finite element approach with Fortran 77 software. —
(Ellis Horwood series in mechanical engineering).
1. Strains and stresses — Data processing
2. FORTRAN (Computer program language)
I. Title
620.1'123'02855133   TA407

Library of Congress Card No. 87–9289

ISBN 0–7458–0246–X (Ellis Horwood Limited — Library Edn.)
ISBN 0–7458–0302–4 (Ellis Horwood Limited — Student Edn.)
ISBN 0–470–20895–3 (Halsted Press)

Printed in Great Britain by R. J. Acford, Chichester

Table of Contents

Preface

The widespread availability of computers has led to profound changes in many branches of engineering. Nowhere are these changes more evident than in the engineering design office where computer-aided techniques are now routinely employed to analyse the stresses in components whose complexity precludes the use of traditional methods. Using these techniques the engineer can set up a computer model of a component prototype and assess its behaviour under any desired system of loading, without recourse to a costly and time-consuming experimental programme.

The finite element method is currently the most popular technique, and numerous commercial software packages are now available for its implementation. Programs that were originally developed by research specialists on multi-million pound mainframe computers are now increasingly available on mini- and micro-computers whose cost is within the reach of even the smallest engineering design office. As a consequence, the engineer is coming under increasing pressure to justify his designs using these programs. Although he will not normally require a detailed understanding of the software, it is essential, however, that he is aware of the assumptions upon which the technique is based, and its inherent limitatations.

To reflect these changes in engineering practice, many undergraduate courses in stress analysis now include an account of the finite element method and its applications. Of the books used as an accompaniment to these courses, many still concentrate upon analytical techniques and include only a brief introduction to the method. Of the numerous books devoted to the method itself, many employ a theoretical treatment which is not well suited to the needs of the less mathematically minded reader.

This book is an attempt to combine in a single volume an outline of analytical techniques and a thorough grounding in the engineering approach to two-dimensional elastic finite element stress analysis. By way of an introduction to the computational techniques involved, a FORTRAN 77 program FIESTA1 is developed for the solution of a simple one-dimensional problem. A more advanced program FIESTA2 for solving two-dimensional problems is later developed, and a range of engineering applications is described. These programs are designed as a teaching aid to familiarise the reader with some of the typical features of a commercial package.

The book is written for students of stress analysis in the fields of mechanical, civil and aeronautical engineering. It is also hoped that practising engineers wishing to update their knowledge of the subject will find the book of value. The reader should have already completed an elementary course in the mechanics of materials, otherwise known as the strength of materials. A basic knowledge of matrix algebra is assumed, and the reader is expected to have some previous experience of computer programming in FORTRAN 77.

A number of problems for solution by the reader are included at the end of most chapters, and a list of references will be found at the end of the book for those wishing to explore the subject further.

COMPUTER SOFTWARE

Copies of the programs FIESTA1 and FIESTA2 on floppy disc are available in FORTRAN 77 for various computers, and in BASIC for the BBC microcomputer. For further information please contact Ellis Horwood Ltd, Market Cross House, Cooper Street, Chichester, West Sussex PO19 1EB, Although these programs have been carefully tested, neither the author nor the publishers can accept any responsibility for any liability whatsoever that might arise through their use.

King's College London   D. N. Fenner
January 1986

List of Symbols

The mathematical symbols used in this book are defined in the following list. A glossary of the FORTRAN variables employed in the computer programs FIESTA1 and FIESTA2 is given in Appendix A1.

A	area of an element
A	area of an inclined plane
A	cross-section area
a	half-crack length
a, b	inner and outer radii of cylinder or disc
$a_i, b_i, c_i,$	... functions of the nodal point coordinates used to define the shape functions for triangular elements
$[B]$	used in the relationship $[e] = [B][\delta^e]$
$[B_c]$	matrix $[B]$ evaluated at element centroid
b	semi-bandwidth of system stiffness matrix
b_i	body force component in the i direction
$[b]$	column matrix of body force components
C_i, D_i, E_i	polynomial coefficients $(i = 1, 2, \ldots)$
c	used as a subscript to denote centroidal value
c, s	cosine and sine of the angle of inclination of the axis of a frame element
$[D]$	matrix of the elastic properties of an element
E	elastic modulus
E^*	equivalent elastic modulus in plane strain
e	used as a subscript or superscript to denote the value for an element e
e_{ii}, γ_{ij}	direct and shear strains for orthogonal axes i, j
e_p	principal strain $(p = 1, 2, 3, \text{max or min})$
e_θ	direct strain in a line segment

$[e]$	strain tensor
$[e]$	column matrix of element strains
F_i^e	force at node i for element e
$[F]$	system matrix of nodal forces
$[F^e]$	element matrix of nodal forces
$[F_{mn}^e]$	column matrix of forces at boundary nodes m, n
G	shear modulus
$\mathcal{G}_I$	potential energy release rate for a cracked body
$\mathcal{G}_{Ic}$	rate of energy dissipation during crack extension
h	plate thickness
I	second moment of area
I_i	invariant of the stress tensor ($i = 1, 2$ or 3)
i	elimination step number in Gaussian elimination
i, j, k	element nodes
J	polar second moment of area of a shaft cross-section
K	ratio of external to internal diameter of a cylinder
K_I	mode I stress intensity factor
K_{Ic}	fracture toughness
K_{rc}	stiffness coefficient in row r, column c of $[K]$
$[K]$	system stiffness matrix
$[K^e]$	element stiffness matrix
l	element length
l, m, n	direction cosines
M	bending moment
M_i^e	bending moment at node i of element e
m, n	nodes defining an element edge on which stresses are prescribed
$[N]$	matrix of element shape functions
$[N_{mn}]$	matrix of shape functions on an element edge m, n
n	number of equations to be solved
n_e	number of elements in a system
n_n	number of nodes in a system
n_r, n_s	number of nodes in r and s directions
P	point load
P_e	element spring force
P_i^e, Q_i^e	components of force at node i of element e
p	pressure
q	intensity of a distributed load
R	radius of Mohr's circle
R	elimination ratio in Gaussian elimination
R	fillet radius
R, θ, ϕ	spherical coordinates
r, c	row, column indices
r, s	local coordinates used in mesh generation
r, θ, z	cylindrical coordinates
S_e	element surface area
s_e	equivalent element spring stiffness

T	temperature
T	torque
$[T]$	transformation matrix
t_i	surface traction component in the i direction
$\bar{t}_i$	prescribed value of t_i
$[t]$	row vector of surface tractions
$[t_{mn}]$	row vector of surface tractions on an element edge m, n
U	elastic strain energy
$\hat{U}$	elastic strain energy density
u, v, w	displacement components in three orthogonal directions
$[u]$	column matrix of displacement components
V	volume
V	potential energy of external forces
V	shear force
W	work done by an external force
w	lateral deflection of a beam
$\bar{x}, \bar{y}$	local coordinates
x', y'	rotated axes
x, y, z	Cartesian coordinates
α	angle of twist per unit length of shaft
α	coefficient of thermal expansion
α	orientation of a frame element
α, β	orientations of principal axes of stress and strain
γ_θ	shear strain measured with respect to two orthogonal line segments
Δ_e	elongation of a spring element
δ	displacement
$[\delta]$	system matrix of nodal displacements
$[\delta^e]$	element matrix of nodal displacements
η	boundary coordinate measured along an element edge
θ_i	rotation at node i
ν	Poisson's ratio
ν^*	equivalent Poisson's ratio in plane strain
Π	potential energy of a system
ρ	density
$\bar{\sigma}$	hydrostatic stress
σ'_d	deviatoric stresses ($d = 1, 2, 3$)
σ_F	fracture stress
σ_{ij}, τ_{ij}	direct and shear stresses for orthogonal axes i, j
$\bar{\sigma}_m, \bar{\tau}_m$	prescribed direct and shear stresses at boundary node m
σ_p	principal stress ($p = 1, 2, 3$, max or min)
σ_Y	yield stress in simple tension
$\bar{\sigma}_{zz}$	axial stress needed to provide the end constraint in a plane strain problem
$\sigma_\theta, \tau_\theta$	direct and shear stresses on an inclined plane
$[\sigma]$	element stresses
$[\sigma]$	stress tensor

$[\sigma']$	stress tensor for rotated axes
τ_{max}	maximum shear stress at a point
Φ	stress function
ϕ	angle of twist of a shaft
ψ	warping function for a noncircular shaft
ψ	angle of rotation of a point on the cross-section of an axisymmetric shaft
Ω	body force potential function
ω	angular velocity
ω_θ	angle of rotation of a line segment
$\underline{\nabla}^2$	harmonic differential operator
∇^4	biharmonic differential operator

To my wife Graciela and children Andrea and Julian.

Chapter 1

Analysis of Stress

1.1 INTRODUCTION

In this chapter, we consider the way in which *external forces* acting on a solid are transmitted through the solid. These forces may be *surface forces* (such as fluid pressure and contact pressure between bodies), which act over the surface of the solid, or *body forces* (such as weight and inertia forces), which are distributed over the volume of the solid.

The transmission of force through a solid entails the generation of *internal forces*. To analyse these forces, we use a *continuum* model of the material in which matter is assumed to be continuously distributed throughout the region occupied by the solid. The atomic structure of the material is ignored and it becomes possible to quantify properties *at a point*. Whilst external forces may be distributed in an arbitrary fashion, the internal forces in a continuum are assumed to be continuous functions of position.

1.2 STATE OF STRESS AT A POINT

Consider a solid of arbitrary shape which is in equilibrium under the action of a set of external forces (see Fig. 1.1). The distribution of the internal forces at any interior point O can be studied by passing a plane through O. The distribution of force required on such a plane to maintain equilibrium of the isolated portion of the solid (see Fig. 1.2(a)) will, in general, be nonuniform. If, however, we take a small area ΔA of the plane surrounding O, we can assume that the distribution of the resultant force ΔF acting on this area is essentially uniform.

The resultant force ΔF can be resolved into the components ΔF_x, ΔF_y, ΔF_z in the three coordinate directions x, y, z (see Fig. 1.2(b)), where the x axis defines the

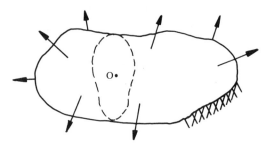

Fig. 1.1—Solid in equilibrium under the action of a set of external forces.

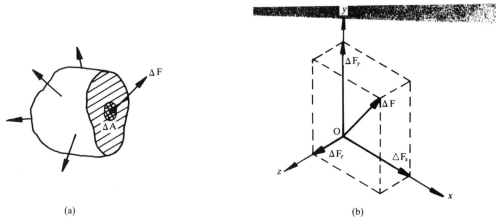

(a)   (b)

Fig. 1.2—(a) Resultant force and (b) Cartesian force components acting on a small portion of an internal plane.

direction of the outward normal to the plane. To measure the intensity of the distribution of each of these forces we define the following three *stress components*:

$$\sigma_{xx} = \lim_{\Delta A \to 0} \left(\frac{\Delta F_x}{\Delta A} \right), \qquad \tau_{xy} = \lim_{\Delta A \to 0} \left(\frac{\Delta F_y}{\Delta A} \right), \qquad \tau_{xz} = \lim_{\Delta A \to 0} \left(\frac{\Delta F_z}{\Delta A} \right)$$

These stresses are the limiting values as the area ΔA is shrunk to the mathematical point O, a limit process which is only meaningful for a continuum. The component σ_{xx} is a *normal* or *direct stress* which measures the intensity of the normal force on the plane at the point O. The components τ_{xy} and τ_{xz} are *shear stresses* which measures the intensity of the shear force on the plane at O. The first of the *double subscripts* used for the stress components indicates the direction of the outward normal to the plane on which the stress acts, and the second indicates the direction of the stress itself.

In a similar way, we could define the stresses at O on any of the infinite number of planes which can be passed through O. It can be shown, however, that the specification of the normal and shear stresses on any *three orthogonal planes* (see Fig. 1.3) is sufficient to define completely the *state of stress* at O. These planes carry the *six* distinct stress components.

$$\sigma_{xx}, \ \sigma_{yy}, \ \sigma_{zz}, \ \tau_{xy}, \ \tau_{yz}, \ \tau_{zx}$$

together with the *complementary shear stresses* $\tau_{yx} = \tau_{xy}$, $\tau_{zy} = \tau_{yz}$ and $\tau_{xz} = \tau_{zx}$.

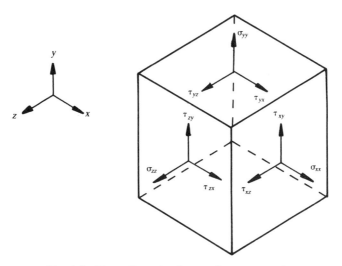

Fig. 1.3—Three-dimensional state of stress at a point.

The *sign convention* for the stresses employed in Fig. 1.3 defines a stress component as positive if the plane on which it acts has an outward normal pointing in the positive (negative) direction of a coordinate axis, and the stress itself also acts in the positive (negative) direction of a coordinate axis. If not, the stress component is taken to be negative. The sign convention for direct stresses is simply that tensile stresses are positive and compressive stresses are negative.

1.3   TWO-DIMENSIONAL STRESS STATE

A two-dimensional or *plane stress state* is one in which $\sigma_{zz} = \tau_{yz} = \tau_{zx} = 0$, and where the remaining stresses are independent of the z coordinate. It is, therefore, completely defined by specifying the normal and shear stresses on any *pair* of orthogonal planes, such as AB and BC in Fig. 1.4, a total of three distinct components σ_{xx}, σ_{yy}, $\tau_{xy}(= \tau_{yx})$ in all.

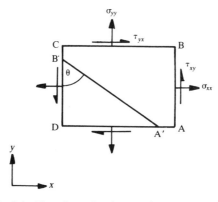

Fig. 1.4—Two-dimensional state of stress at a point.

1.3.1 Stresses on an inclined plane

Next we calculate the normal and shear stresses σ_θ and τ_θ on a plane $A'B'$ inclined at an angle θ to the y direction (see Figs 1.4 and 1.5(a)). Denoting the components of *surface traction* per unit area on $A'B'$ by t_x and t_y (see Fig. 1.5(b)), we can write the equations of force equilibrium for the element $A'B'D$ in the x and y directions as

$$t_x A = \sigma_{xx} Al + \tau_{yx} Am$$
$$t_y A = \tau_{xy} Al + \sigma_{yy} Am$$
(1.1)

where A is the area of the inclined plane $A'B'$. $l = \cos\theta$ and $m = \cos(90° - \theta)$ are the *direction cosines* of the outward normal to $A'B'$ with the x and y directions and

$$l^2 + m^2 = 1$$

Equations (1.1) can be written in *matrix notation* as

$$[t_x \ \ t_y] = [l \ \ m] \begin{bmatrix} \sigma_{xx} & \tau_{xy} \\ \tau_{yx} & \sigma_{yy} \end{bmatrix}$$
(1.2)

or

$$[t] = [l \ \ m][\sigma]$$

where $[\sigma]$ is the *stress tensor*, a 2 by 2 *symmetric* matrix defining the state of stress in the (x, y) coordinate system.

The stresses σ_θ and τ_θ may be expressed in terms of t_x and t_y by

$$\sigma_\theta = t_x l + t_y m, \qquad \tau_\theta = t_x l' + t_y m'$$

or

$$[\sigma_\theta \ \ \tau_\theta] = [t_x \ \ t_y] \begin{bmatrix} l & l' \\ m & m' \end{bmatrix}$$
(1.3)

$$= [t][T]$$

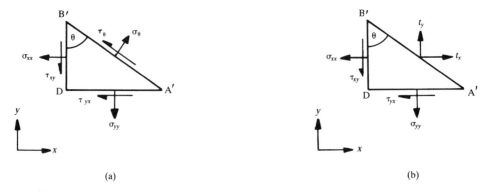

(a)   (b)

Fig. 1.5—(a) Normal and shear stresses and (b) surface tractions acting on an inclined plane.

where $[T]$ is a *transformation matrix* in which $l' = \cos(\theta + 90°), m' = \cos\theta$ are the direction cosines of the tangent to $A'B'$. By substituting for $[t]$ from (1.2) into (1.3), we obtain

$$[\sigma_\theta \quad \tau_\theta] = [l \quad m][\sigma] \begin{bmatrix} l & l' \\ m & m' \end{bmatrix} \tag{1.4}$$

After carrying out the matrix multiplication in (1.4), substituting for the direction cosines and using the trigonometrical identities for $\cos 2\theta$ and $\sin 2\theta$ in terms of $\cos\theta$ and $\sin\theta$ we arrive at

$$\sigma_\theta = \tfrac{1}{2}(\sigma_{xx} + \sigma_{yy}) + \tfrac{1}{2}(\sigma_{xx} - \sigma_{yy})\cos 2\theta + \tau_{xy}\sin 2\theta \tag{1.5}$$
$$\tau_\theta = -\tfrac{1}{2}(\sigma_{xx} - \sigma_{yy})\sin 2\theta + \tau_{xy}\cos 2\theta$$

1.3.2 Mohr's circle of stress

It is convenient to employ a graphical representation of equations (1.5), known as *Mohr's circle of stress*, to display the variations in the normal and shear stresses with the angle θ of the plane $A'B'$. With the aid of the identities

$$p\sin 2\theta + q\cos 2\theta = R\cos(2\theta - 2\alpha)$$
$$p\cos 2\theta - q\sin 2\theta = -R\sin(2\theta - 2\alpha)$$

in which $2\alpha = \tan^{-1}(p/q)$ and $R = \sqrt{p^2 + q^2}$, we can re-express these equations as

$$\sigma_\theta = \tfrac{1}{2}(\sigma_{xx} + \sigma_{yy}) + R\cos(2\theta - 2\alpha) \tag{1.6}$$
$$\tau_\theta = -R\sin(2\theta - 2\alpha)$$

where

$$2\alpha = \tan^{-1}\left(\frac{2\tau_{xy}}{\sigma_{xx} - \sigma_{yy}}\right) \quad\text{and}\quad R = \tfrac{1}{2}\sqrt{(\sigma_{xx} - \sigma_{yy})^2 + 4\tau_{xy}^2}$$

It follows that $(\sigma_\theta, \tau_\theta)$ provide the coordinates of a point $A'B'$ which describes a circle in an anti-clockwise manner as the angle 2θ varies from $0°$ to $360°$ (see Fig. 1.6(a)). Mohr's circle has a radius R, and the centre C has coordinates $(\tfrac{1}{2}(\sigma_{xx} + \sigma_{yy}), 0)$. The points AB and BC, which are $180°$ apart on the circle, have coordinates corresponding to the normal and shear stresses on the planes AB and BC, which are at $90°$ to one another (see Fig. 1.6(b)). Note the minus sign for the τ_θ coordinate of the point BC which results from the difference in the sign conventions for τ_{yx} and τ_θ on the plane BC.

1.3.3 Principal stresses and maximum shear stress

As the plane $A'B'$ is rotated anti-clockwise from the initial inclination $\theta = 0°$ (the point AB in Fig. 1.6(a)), the following special cases arise.

(1) The two orthogonal planes, $\theta = \alpha$ and $\theta = \alpha + 90°$, corresponding to the points A_1B_1 and B_1C_1 on Mohr's circle, carry the maximum and minimum

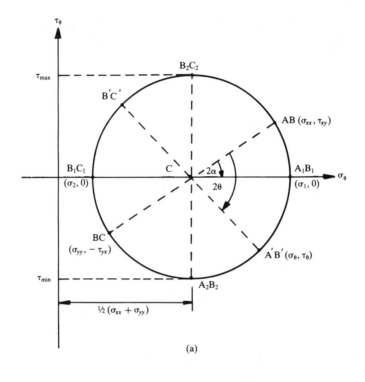

(a)

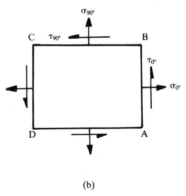

(b)

Fig. 1.6—(a) Mohr's circle for a two-dimensional state of stress and (b) stress components acting on the faces of a rectangular element.

direct stresses, respectively. These are known as the *principal stresses* and are given by

$$\sigma_{1,2} = \tfrac{1}{2}(\sigma_{xx} + \sigma_{yy}) \pm \tfrac{1}{2}\sqrt{(\sigma_{xx} - \sigma_{yy})^2 + 4\tau_{xy}^2} \qquad (1.7)$$

The planes are the *principal planes* and carry no shear stress. The directions of the outward normals to A_1B_1 and B_1C_1 define the *principal axes* x_1, y_1 (see Fig. 1.7(a)).

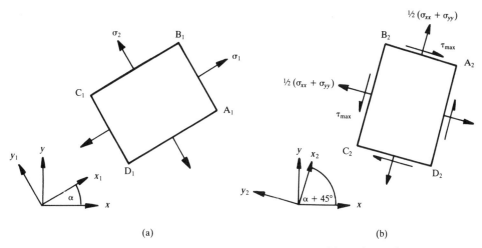

Fig. 1.7—Rotated elements carrying (a) principal stresses and (b) maximum shear stress.

(2)   The two orthogonal planes, $\theta = \alpha + 135°$ and $\theta = \alpha + 45°$, corresponding to
      the points B_2C_2 and A_2B_2 on Mohr's circle, carry the *maximum and minimum
      shear stresses*, respectively. These are given by

$$\tau_{max} = -\tau_{min} = R = \tfrac{1}{2}(\sigma_1 - \sigma_2) \tag{1.8}$$

The directions of the outward normals to these planes define the set of axes
x_2, y_2 (see Fig. 1.7(b)) which are oriented at $45°$ to the principal axes. The
value of τ_{max} given by (1.8) is not necessarily the maximum shear stress at the
point, as we shall see later.

1.3.4 Rotation of axes
Our next task is to obtain the transformation rule which enables the state of stress to
be redefined in terms of new axes x', y' obtained by rotating the axes x, y
anti-clockwise through an angle θ (see Fig. 1.8). We can identify the normal and

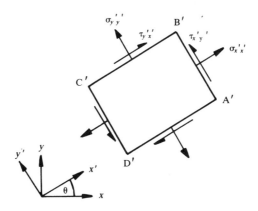

Fig. 1.8—Two-dimensional stress state defined in terms of rotated axes.

shear stresses on the plane A'B' as $\sigma_{x'x'} = \sigma_\theta$, $\tau_{x'y'} = \tau_\theta$, which are given by (1.4) as

$$[\sigma_{x'x'} \quad \tau_{x'y'}] = [l \quad m][\sigma][T] \tag{1.9}$$

where l, m denote the direction cosines of the axis x'. On the orthogonal plane B'C' the corresponding stresses $\sigma_{y'y'} = \sigma_{\theta+90°}$, $\tau_{y'x'} = -\tau_{\theta+90°}$ are given by a similar matrix equation

$$[\tau_{y'x'} \quad \sigma_{y'y'}] = [l' \quad m'][\sigma][T] \tag{1.10}$$

where l', m' are the direction cosines of the axis y'. Equations (1.9) and (1.10) are combined to give the required transformation rule

$$\begin{bmatrix} \sigma_{x'x'} & \tau_{x'y'} \\ \tau_{y'x'} & \sigma_{y'y'} \end{bmatrix} = \begin{bmatrix} l & m \\ l' & m' \end{bmatrix} \begin{bmatrix} \sigma_{xx} & \tau_{xy} \\ \tau_{yx} & \sigma_{yy} \end{bmatrix} \begin{bmatrix} l & l' \\ m & m' \end{bmatrix}$$

or simply

$$[\sigma'] = [T]^T[\sigma][T] \tag{1.11}$$

where $[\sigma']$ is the stress tensor defining the stress state in terms of the new axes x', y'.

This transformation may also be performed graphically using Mohr's stress circle (see Fig. 1.6(a)) by rotating the diameter joining the points AB and BC clockwise through an angle 2θ. The points A'B' and B'C' thereby located have coordinates $(\sigma_{x'x'}, \tau_{x'y'})$ and $(\sigma_{y'y'}, -\tau_{y'x'})$, respectively.

An alternative method for the calculation of principal stresses, which we shall later generalise for a three-dimensional stress state, involves the manipulation of the stress tensor $[\sigma]$. If A'B' is a principal plane (see Fig. 1.9) carrying a principal stress σ_p, the shear stress on the plane is zero and according to (1.1) the surface tractions are $t_x = \sigma_p l$ and $t_y = \sigma_p m$, where l, m are the direction cosines of the principal axis x'. These conditions can be expressed by the matrix equation

$$\begin{bmatrix} t_x \\ t_y \end{bmatrix} = \begin{bmatrix} \sigma_p & 0 \\ 0 & \sigma_p \end{bmatrix} \begin{bmatrix} l \\ m \end{bmatrix} = \sigma_p[I] \begin{bmatrix} l \\ m \end{bmatrix} \tag{1.12}$$

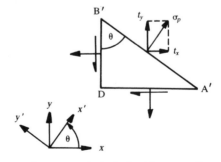

Fig. 1.9—Surface tractions on a principal plane for a two-dimensional stress state.

where

$$[I] = \begin{bmatrix} 1 & 0 \\ 0 & 1 \end{bmatrix}$$

is the 2 by 2 *unit matrix*. Taking the transposes of both sides of (1.2), we obtain

$$\begin{bmatrix} t_x \\ t_y \end{bmatrix} = [\sigma]^{\mathrm{T}} \begin{bmatrix} l \\ m \end{bmatrix} = [\sigma] \begin{bmatrix} l \\ m \end{bmatrix} \qquad (1.13)$$

and equating the right-hand sides of (1.12) and (1.13) leads to

$$([\sigma] - \sigma_p[I]) \begin{bmatrix} l \\ m \end{bmatrix} = 0 \qquad (1.14)$$

a pair of homogeneous algebraic equations in l and m. For (1.14) to have *non-trivial* solutions, it follows that the determinant of the coefficient matrix must be zero, or

$$|[\sigma] - \sigma_p[I]| = 0 \qquad (1.15)$$

This requires that

$$\begin{bmatrix} \sigma_{xx} - \sigma_p & \tau_{xy} \\ \tau_{yx} & \sigma_{yy} - \sigma_p \end{bmatrix} = 0 \qquad (1.16)$$

which we can expand to give the *quadratic* equation

$$\sigma_p^2 - I_1\sigma_p + I_2 = 0 \qquad (1.17)$$

where $I_1 = \sigma_{xx} + \sigma_{yy}$ and $I_2 = \sigma_{xx}\sigma_{yy} - \tau_{xy}^2$. The two roots of (1.17) are the principal stresses, and for each root there is a solution for the direction cosines l, m defining the direction of the corresponding principal axis. The principal stresses are seen to be the *eigenvalues* of the stress tensor $[\sigma]$, whilst the direction cosines are the *eigenvectors* of $[\sigma]$.

The magnitudes of the principal stresses must be independent of the orientation of the original axes x, y used to define the stress state. This will only be the case if the coefficients I_1 and I_2 of the quadratic equation (1.17) are independent of this orientation. For this reason, I_1 and I_2 are known as the first and second *invariants* of the stress tensor.

1.4 THREE-DIMENSIONAL STRESS STATE

In the case of a general three-dimensional state of stress an inclined plane $A'B'C'$ (see Fig. 1.10) has three components of surface traction t_x, t_y, t_z acting on it. These are obtained from the three-dimensional form of (1.2) given by

$$[t_x \quad t_y \quad t_x] = [l \quad m \quad n][\sigma] \qquad (1.18)$$

where l, m, n are the direction cosines of the outward normal to $A'B'C'$ with

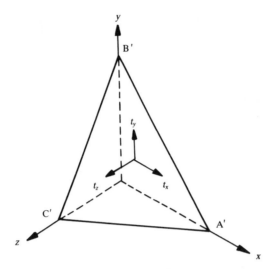

Fig. 1.10—Surface tractions on an inclined plane for a three-dimensional stress state.

respect to the x, y and z axes, respectively. The stress tensor is now

$$[\sigma] = \begin{bmatrix} \sigma_{xx} & \tau_{xy} & \tau_{xz} \\ \tau_{yx} & \sigma_{yy} & \tau_{yz} \\ \tau_{zx} & \tau_{zy} & \sigma_{zz} \end{bmatrix} \tag{1.19}$$

and in the transformation rule for $[\sigma]$ given by (1.11) the transformation matrix becomes

$$[T] = \begin{bmatrix} l & l' & l'' \\ m & m' & m'' \\ n & n' & n'' \end{bmatrix} \tag{1.20}$$

where l'', m'' and n'' are the direction cosines of the z' axis.

Principal stresses, corresponding to the eigenvalues of $[\sigma]$, are given by the three-dimensional form of (1.16) given by

$$\begin{bmatrix} \sigma_{xx} - \sigma_p & \tau_{xy} & \tau_{xz} \\ \tau_{yx} & \sigma_{yy} - \sigma_p & \tau_{yz} \\ \tau_{zx} & \tau_{zy} & \sigma_{zz} - \sigma_p \end{bmatrix} = 0 \tag{1.21}$$

which leads to the *cubic* equation

$$\sigma_p^3 - I_1\sigma_p^2 + I_2\sigma_p - I_3 = 0 \tag{1.22}$$

where the three stress invariants are

$$I_1 = \sigma_{xx} + \sigma_{yy} + \sigma_{zz}$$

$$I_2 = \sigma_{xx}\sigma_{yy} + \sigma_{yy}\sigma_{zz} + \sigma_{zz}\sigma_{xx} - \tau_{xy}^2 - \tau_{yz}^2 - \tau_{zx}^2$$

$$I_3 = \sigma_{xx}\sigma_{yy}\sigma_{zz} + 2\tau_{xy}\tau_{yz}\tau_{zx} - \sigma_{xx}\tau_{yz}^2 - \sigma_{yy}\tau_{zx}^2 - \sigma_{zz}\tau_{xy}^2$$

It follows that there are three principal stresses, three principal orthogonal planes and three principal axes for a three-dimensional stress state.

It is possible to extend Mohr's circle representation to three dimensions if we start with a definition of the state of stress in terms of the principal stresses σ_1, σ_2, σ_3 (see Fig. 1.11(a)). For a plane A'B' parallel to the third principal axis z_1 (see Fig. 1.11(b)) the normal and shear stresses are given by (1.5), with $\sigma_{xx} = \sigma_1$, $\sigma_{yy} = \sigma_2$, $\tau_{xy} = 0$ and $\theta = \theta_3$, as

$$\sigma_{\theta_3} = \tfrac{1}{2}(\sigma_1 + \sigma_2) + \tfrac{1}{2}(\sigma_1 - \sigma_2)\cos 2\theta_3$$

$$\tau_{\theta_3} = -\tfrac{1}{2}(\sigma_1 - \sigma_2)\sin 2\theta_3$$

Similar pairs of equations can be derived for the stresses on planes parallel to the first principal axis x_1, and second principal axis y_1. Each of these pairs of equations has a Mohr's circle representation, as shown in Fig. 1.12. Although these circles only represent special sets of planes, it can be shown that the normal and shear stresses for *all possible* planes give the coordinates of points which lie within the shaded region *between* the three circles. This being the case, it follows that the maximum shear stress is given by

$$\tau_{\max} = \tfrac{1}{2}\left|\sigma_1 - \sigma_2\right|, \quad \tfrac{1}{2}\left|\sigma_2 - \sigma_3\right| \quad \text{or} \quad \tfrac{1}{2}\left|\sigma_3 - \sigma_1\right|$$

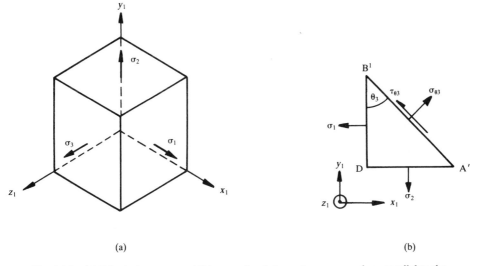

(a)   (b)

Fig. 1.11—(a) Principal stresses and (b) normal and shear stresses on a plane parallel to the third principal axis, for a three-dimensional stress state.

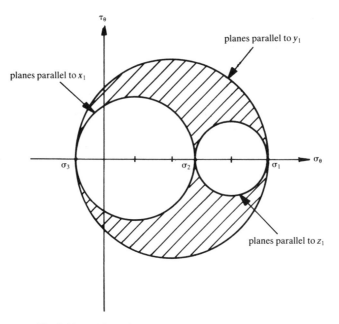

Fig. 1.12—Mohr's circles for a three-dimensional stress state.

whichever is the greatest. In other words,

$$\tau_{max} = \tfrac{1}{2}(\sigma_{max} - \sigma_{min}) \tag{1.23}$$

where σ_{max} is the maximum principal stress and σ_{min} is the minimum principal stress.

It is important to notice that for a two-dimensional stress state, where $\sigma_3 = 0$, the maximum shear stress on any possible plane through the point under consideration is only given by (1.8) in the special case where σ_1 and σ_2 are the maximum and minimum principal stresses, respectively.

1.5 STRESS EQUILIBRIUM

So far in this chapter, we have been concerned with the definition and analysis of the state of stress at a point in a solid. The question now arises as to how the variation in the stress state from one point to another in a solid can be described.

1.5.1 Differential equations of equilibrium

When a solid is in equilibrium under the action of a set of external forces, it follows that any isolated part of the solid must also be in equilibrium. The small rectangular element with sides of length Δx and Δy shown in Fig. 1.13 is such a part of a solid of unit thickness in which the stress state is assumed to be two dimensional. There are body forces present which have components b_x and b_y per unit volume in the x and y directions, respectively.

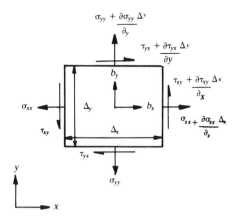

Fig. 1.13—Stress components and body forces acting on a small rectangular element.

In general, the stress components will be functions of both x and y. It follows that, if the average direct stress on the left-hand face of the element is σ_{xx}, the average direct stress on the right-hand face is greater by an amount $(\partial\sigma_{xx}/\partial x)\Delta x$. This increment accounts for the variation in σ_{xx} over the distance Δx separating the two faces. The other stresses increment in a similar manner. For force equilibrium in the x direction, we require

$$\sigma_{xx}\,\Delta y + \tau_{yx}\,\Delta x = \left(\sigma_{xx} + \frac{\partial\sigma_{xx}}{\partial x}\,\Delta x\right)\Delta y$$

$$-\left(\tau_{yx} + \frac{\partial\tau_{yx}}{\partial y}\,\Delta y\right)\Delta x + b_x\,\Delta x\,\Delta y = 0$$

which simplifies to

$$\frac{\partial\sigma_{xx}}{\partial x} + \frac{\partial\tau_{xy}}{\partial y} + b_x = 0 \tag{1.24a}$$

Similarly, we can establish the condition for force equilibrium in the y direction as

$$\frac{\partial\tau_{yx}}{\partial x} + \frac{\partial\sigma_{yy}}{\partial y} + b_y = 0 \tag{1.24b}$$

Finally, we can show that moment equilibrium about any axis normal to the xy plane simply confirms the complementary shear stress condition $\tau_{yx} = \tau_{xy}$.

By a similar means, we can derive the equations of equilibrium for a three-dimensional stress state. In this case, there are a total of *six* equilibrium conditions: there are three for moment equilibrium about each of three axes x, y and z which confirm that $\tau_{yx} = \tau_{xy}$, $\tau_{zy} = \tau_{yz}$ and $\tau_{xz} = \tau_{zx}$; the other three are for force equilibrium in the three directions x, y and z which require that

$$\frac{\partial \sigma_{xx}}{\partial x} + \frac{\partial \tau_{xy}}{\partial y} + \frac{\partial \tau_{xz}}{\partial z} + b_x = 0$$

$$\frac{\partial \tau_{yx}}{\partial x} + \frac{\partial \sigma_{yy}}{\partial y} + \frac{\partial \tau_{yz}}{\partial z} + b_y = 0 \qquad (1.25)$$

$$\frac{\partial \tau_{zx}}{\partial x} + \frac{\partial \tau_{zy}}{\partial y} + \frac{\partial \sigma_{zz}}{\partial z} + b_z = 0$$

where b_z is the z direction component of body force.

It is clear that we have only the three differential equations of equilibrium (1.25) in the total of six unknown stress components (or the two equations (1.24) in three stresses for a two-dimensional problem). In general, therefore, the problem of solving for the stresses is *statically indeterminate* and we must obtain additional equations using considerations other than equilibrium.

1.5.2 Stress boundary conditions

The equations of equilibrium derived above must be satisfied at every point in the interior of a solid. We turn now to the conditions which must be satisfied on the boundary of the solid. These may be expressed in terms of either stresses or displacements, or a combination of the two.

It is convenient to define stress boundary conditions in terms of the prescribed values $\bar{t}_x, \bar{t}_y$ and $\bar{t}_z$ for the surface tractions. In this way, infinitesimal portions of the surface can be treated as the inclined planes $A'B'C'$ of section 1.4 (see Fig. 1.10). It follows from (1.18) that the solution for the stress state $[\sigma]$ must satisfy

$$\begin{bmatrix} t_x & t_y & t_x \end{bmatrix} = \begin{bmatrix} l & m & n \end{bmatrix}[\sigma]$$
$$= \begin{bmatrix} \bar{t}_x & \bar{t}_y & \bar{t}_z \end{bmatrix} \qquad (1.26)$$

at every point on the boundary where stresses are prescribed.

1.6 USE OF CYLINDRICAL COORDINATES

Up to this point, we have employed only Cartesian coordinates in the analysis of stress. In engineering practice, it is often convenient to employ other orthogonal coordinate systems to suit the particular application.

A wide range of engineering components (such as cylinders, shafts and rotors) has an axisymmetric geometry which is easiest to describe using *cylindrical coordinates* in which the position of an arbitrary point P is defined by r, θ and z (see Fig. 1.14(a)). The three orthogonal planes chosen for the purpose of defining the stress state have outward normals in the r, θ and z directions (see Fig. 1.14(b)). The

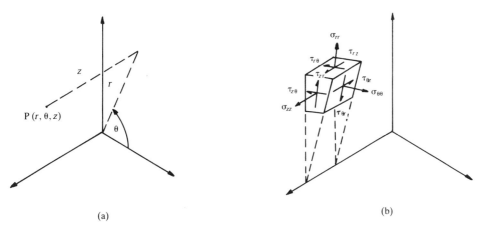

Fig. 1.14—(a) Cylindrical coordinate system and (b) the corresponding state of stress at a point.

corresponding stress tensor

$$[\sigma'] = \begin{bmatrix} \sigma_{rr} & \tau_{r\theta} & \tau_{rz} \\ \tau_{\theta r} & \sigma_{\theta\theta} & \tau_{\theta z} \\ \tau_{zr} & \tau_{z\theta} & \sigma_{zz} \end{bmatrix} \qquad (1.27)$$

can be expressed in terms of $[\sigma]$, defined in Cartesian coordinates, with the aid of the transformation rule (1.11). This gives

$$[\sigma'] = [T]^{\mathrm{T}}[\sigma][T]$$

where

$$[T] = \begin{bmatrix} l & l' & l'' \\ m & m' & m'' \\ n & n' & n'' \end{bmatrix} = \begin{bmatrix} \cos\theta & -\sin\theta & 0 \\ \sin\theta & \cos\theta & 0 \\ 0 & 0 & 1 \end{bmatrix} \qquad (1.28)$$

The three direct stresses σ_{rr}, $\sigma_{\theta\theta}$ and σ_{zz} are often referred to as the *radial*, *hoop* or *circumferential* and *axial stresses*, respectively.

The three principal stresses are given by the roots of the cubic equation (1.22) with the stress invariants now defined as

$$I_1 = \sigma_{rr} + \sigma_{\theta\theta} + \sigma_{zz}$$

$$I_2 = \sigma_{rr}\sigma_{\theta\theta} + \sigma_{\theta\theta}\sigma_{zz} + \sigma_{zz}\sigma_{rr} - \tau_{r\theta}^2 - \tau_{\theta z}^2 - \tau_{zr}^2 \qquad (1.29)$$

$$I_3 = \sigma_{rr}\sigma_{\theta\theta}\sigma_{zz} + 2\tau_{r\theta}\tau_{\theta z}\tau_{zr} - \sigma_{rr}\tau_{\theta z}^2 - \sigma_{\theta\theta}\tau_{zr}^2 - \sigma_{zz}\tau_{r\theta}^2$$

The derivation of the differential equations of stress equilibrium in cylindrical coordinates is similar to that for Cartesian coordinates described in section 1.5.1. If the forces on an infinitesimal solid element satisfy equilibrium in the r, θ and z

directions, it can be shown that

$$\frac{\partial \sigma_{rr}}{\partial r} + \frac{1}{r}\frac{\partial \tau_{r\theta}}{\partial \theta} + \frac{\partial \tau_{rz}}{\partial z} + \frac{\sigma_{rr} - \sigma_{\theta\theta}}{r} + b_r = 0$$

$$\frac{\partial \tau_{\theta r}}{\partial r} + \frac{1}{r}\frac{\partial \sigma_{\theta\theta}}{\partial \theta} + \frac{\partial \tau_{\theta z}}{\partial z} + \frac{2\tau_{r\theta}}{r} + b_\theta = 0 \qquad (1.30)$$

$$\frac{\partial \tau_{zr}}{\partial r} + \frac{1}{r}\frac{\partial \tau_{z\theta}}{\partial \theta} + \frac{\partial \sigma_{zz}}{\partial z} + \frac{\tau_{rz}}{r} + b_z = 0$$

where b_r, b_θ and b_z are the components of body force per unit volume in the r, θ and z directions, respectively.

Stress boundary conditions are expressed in terms of prescribed values $\bar{t}_r$, $\bar{t}_\theta$ and $\bar{t}_z$ for the components of surface traction in the r, θ and z directions. In place of (1.26), we now have

$$[t_r \quad t_\theta \quad t_z] = [l \quad m \quad n][\sigma]$$
$$= [\bar{t}_r \quad \bar{t}_\theta \quad \bar{t}_z] \qquad (1.31)$$

for points on the boundary where stresses are prescribed.

PROBLEMS

1.1   Find the principal stresses, the orientation of the principal axes of stress, and the maximum shear stress for the following cases of plane stress.

(i) $\sigma_{xx} = \quad 40$, $\sigma_{yy} = \quad\quad 0$, $\tau_{xy} = \quad 80$ MN/m^2
(ii) $\sigma_{xx} = 140$, $\sigma_{yy} = \quad 20$, $\tau_{xy} = -60$ MN/m^2.
(iii) $\sigma_{xx} = 120$, $\sigma_{yy} = \quad 50$, $\tau_{xy} = \quad 40$ MN/m^2.
(iv) $\sigma_{rr} = -40$, $\sigma_{\theta\theta} = -100$, $\tau_{r\theta} = -50$ MN/m^2.

(Answers:
(i)   102.5 MN/m^2, − 62.5 MN/m^2;   38.0°, 128.0°; 82.5 MN/m^2.
(ii)   164.9 MN/m^2, − 4.9 MN/m^2; −22.5°,   67.5°; 84.9 MN/m^2.
(iii)   138.2 MN/m^2,   31.8 MN/m^2;   24.4°, 114.4°; 69.1 MN/m^2.
(iv)   −11.7 MN/m^2, −128.3 MN/m^2; −29.5°,   60.5°; 64.2 MN/m^2.)

1.2   A state of plane stress at point in a solid is defined by $\sigma_{xx} = 10$ MN/m^2, $\sigma_{yy} = -20$ MN/m^2 and $\tau_{xy} = 5$ MN/m^2. Find the orientations of the planes on which the direct stress is equal to zero, and the magnitudes of the shear stresses on these planes.

(Answers: 45°, −26.6°; ±15 MN/m^2.)

1.3   Find the principal stresses, the direction cosines of the principal axes and the maximum shear stress for the three-dimensional stress state defined by

$$[\sigma] = \begin{bmatrix} 30 & 10 & 10 \\ 10 & 0 & 20 \\ 10 & 20 & 0 \end{bmatrix} \quad \text{MN/m}^2.$$

(Answers: 40 MN/m^2, 10 MN/m^2, -20 MN/m^2; $2/\sqrt{6}$, $1/\sqrt{6}$, $1/\sqrt{6}$; $-1/\sqrt{3}$, $1/\sqrt{3}$, $1/\sqrt{3}$; 0, $-1/\sqrt{2}$, $1/\sqrt{2}$; 30 MN/m^2.)

1.4   Show that the normal and shear stresses on the octahedral plane equally inclined to the three principal axes of a three-dimensional stress state are given by

$$\sigma_{oct} = \tfrac{1}{3}(\sigma_1 + \sigma_2 + \sigma_3)$$

and

$$\tau_{oct} = \tfrac{1}{3}\sqrt{(\sigma_1 - \sigma_2)^2 + (\sigma_2 - \sigma_3)^2 + (\sigma_3 - \sigma_1)^2}$$

1.5   A cantilever beam of rectangular cross-section, breadth b and depth $2c$ carries a uniformly distributed load of intensity q per unit length over its upper surface. The stress state may be taken as two dimensional, and the solution for σ_{xx} is given by elementary beam theory as My/I_{zz}. By finding solutions for the equations of stress equilibrium which satisfy the boundary conditions on the upper and lower surfaces of the beam show that

$$\sigma_{yy} = \frac{-q}{6I_{zz}}(y^3 - 3c^2y + 2c^3), \qquad \tau_{xy} = \frac{-qx}{2I_{zz}}(c^2 - y^2)$$

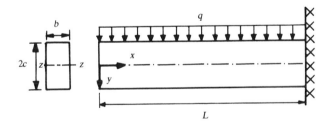

Chapter 2

Analysis of Strain

2.1 INTRODUCTION

When a solid body is acted on by external forces, the changes in its configuration can be defined in terms of the *displacements* of each point in the body. Two types of displacement are possible: *rigid-body displacement* of the body as a whole consisting of combined translation and rotation (see Fig. 2.1), and *deformation* consisting of displacements of points within the body relative to one another (see Fig. 2.2). Of these it is the deformation which is associated with the presence of internal stresses and which we examine in this chapter.

2.2 STATE OF STRAIN AT A POINT

The analysis of deformation involves only geometric considerations and requires no assumptions regarding the material, except that it can be adequately modelled as a continuum. For simplicity, we start by restricting our attention to the two-dimensional or *plane strain* deformation of a solid in which the displacement of any point is defined by the components u and v in the xy plane.

A point O with coordinates (x, y) in an undeformed solid moves to the new position O' with coordinates (x', y') in the deformed solid (see Fig. 2.3) where

$$x' = f(x, y) \quad \text{and} \quad y' = g(x, y) \tag{2.1}$$

Assuming that during deformation no holes or overlaps develop in the solid, we require f and g to be *continuous* and *single-valued functions* of x and y. Equations (2.1) can be written in terms of the displacements u and v as

$$x' = x + u \quad \text{and} \quad y' = y + v \tag{2.2}$$

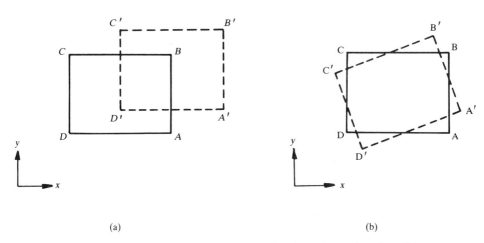

(a)   (b)

Fig. 2.1.—(a) Rigid-body translation and (b) rigid-body rotation of a solid.

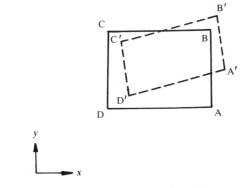

Fig. 2.2—Deformation of a solid.

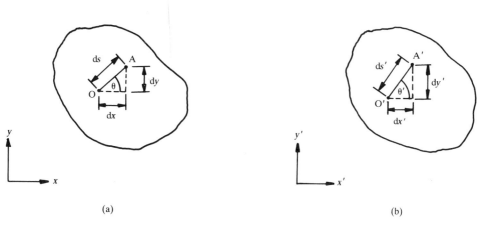

(a)   (b)

Fig. 2.3—(a) Undeformed geometry and (b) deformed geometry of a line segment used to
define the direct strain.

The deformation in the vicinity of O can be analysed if we examine a line segment OA of length ds between O and the adjacent point A with coordinates $(x + dx, y + dy)$. In the deformed body A moves to A′ with coordinates $(x′ + dx′, y′ + dy′)$ and the line segment O′A′ now has a length $ds′$. Using (2.1) and (2.2), we can write $dx′$ and $dy′$ as the total differentials

$$dx' = \frac{\partial x'}{\partial x}\,dx + \frac{\partial x'}{\partial y}\,dy = \left(1 + \frac{\partial u}{\partial x}\right)dx + \frac{\partial u}{\partial y}\,dy$$

$$dy' = \frac{\partial y'}{\partial x}\,dx + \frac{\partial y'}{\partial y}\,dy = \frac{\partial v}{\partial x}\,dx + \left(1 + \frac{\partial v}{\partial y}\right)dy$$

$$(2.3)$$

The *direct strain* in the line segment O′A′ is defined as

$$e_\theta = \frac{ds' - ds}{ds} \tag{2.4}$$

but, rather than calculate e_θ directly, it is mathematically simpler first to find the quantity

$$X = \frac{1}{2}\left(\frac{ds'^2 - ds^2}{ds^2}\right) = e_\theta + \tfrac{1}{2}e_\theta^2 \tag{2.5}$$

where

$$ds^2 = dx^2 + dy^2 \quad \text{and} \quad ds'^2 = dx'^2 + dy'^2$$

Substituting for $dx′$ and $dy′$ from (2.3) into (2.5) and putting

$$\frac{dx}{ds} = \cos\theta \quad \text{and} \quad \frac{dy}{ds} = \sin\theta \tag{2.6}$$

we obtain

$$X = \left\{\frac{\partial u}{\partial x} + \frac{1}{2}\left[\left(\frac{\partial u}{\partial x}\right)^2 + \left(\frac{\partial v}{\partial x}\right)^2\right]\right\}\cos^2\theta$$

$$+ \left\{\frac{\partial v}{\partial y} + \frac{1}{2}\left[\left(\frac{\partial u}{\partial y}\right)^2 + \left(\frac{\partial v}{\partial y}\right)^2\right]\right\}\sin^2\theta$$

$$+ \left(\frac{\partial u}{\partial y} + \frac{\partial v}{\partial x} + \frac{\partial u}{\partial x}\frac{\partial u}{\partial y} + \frac{\partial v}{\partial x}\frac{\partial v}{\partial y}\right)\sin\theta\cos\theta$$

$$= e_\theta + \tfrac{1}{2}e_\theta^2$$

$$(2.7)$$

At this point, we assume that the *strains are small* and that the quadratic terms in (2.7) can be discarded to give

$$e_\theta = \frac{\partial u}{\partial x}\cos^2\theta + \frac{\partial v}{\partial y}\sin^2\theta + \left(\frac{\partial u}{\partial y} + \frac{\partial v}{\partial x}\right)\sin\theta\cos\theta \tag{2.8}$$

As a result of the deformation the line segment OA *rotates* anti-clockwise through an angle $\omega_\theta = \theta' - \theta$ (see Fig. 2.4). Provided that this angle is *small*, it can be written as

$$\omega_\theta = \sin(\theta' - \theta)$$
$$= \sin\theta' \cos\theta - \cos\theta' \sin\theta \qquad (2.9)$$
$$= \frac{dy'}{ds'} \cos\theta - \frac{dx'}{ds'} \sin\theta$$

The total derivatives in (2.9) are obtained from (2.3) as

$$\frac{dx'}{ds'} = \left(1 + \frac{\partial u}{\partial x}\right)\frac{dx}{ds'} + \frac{\partial u}{\partial y}\frac{dy}{ds'}$$
$$\frac{dy'}{ds'} = \frac{\partial v}{\partial x}\frac{dx}{ds'} + \left(1 + \frac{\partial v}{\partial y}\right)\frac{dy}{ds'} \qquad (2.10)$$

where, with the aid of (2.4) and (2.6), we can show that

$$\frac{dx}{ds'} = \frac{dx}{ds}\frac{ds}{ds'} = \frac{\cos\theta}{1 + e_\theta} \approx \cos\theta$$

$$\frac{dy}{ds'} = \frac{dy}{ds}\frac{ds}{ds'} = \frac{\sin\theta}{1 + e_\theta} \approx \sin\theta$$

Substituting (2.10) into (2.9), we obtain

$$\omega_\theta = -\frac{\partial u}{\partial x}\cos\theta\sin\theta + \frac{\partial v}{\partial y}\cos\theta\sin\theta + \frac{\partial v}{\partial x}\cos^2\theta - \frac{\partial u}{\partial y}\sin^2\theta \qquad (2.11)$$

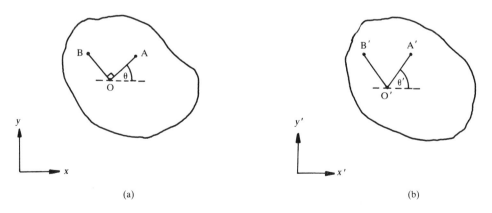

(a)   (b)

Fig. 2.4—(a) Undeformed geometry and (b) deformed geometry of a pair of orthogonal line segments used to define the shear strain.

For a line segment OB at right angles to OA (see Fig. 2.4) the corresponding rotation is found by replacing θ with $\theta + 90°$ in (2.11) to give

$$\omega_{\theta+90°} = \frac{\partial u}{\partial x}\cos\theta\sin\theta - \frac{\partial v}{\partial y}\cos\theta\sin\theta + \frac{\partial v}{\partial x}\sin^2\theta - \frac{\partial u}{\partial y}\cos^2\theta \quad (2.12)$$

It follows that the *shear strain* γ_θ, measured by the reduction in the angle between the orthogonal line segments OA and OB during deformation, is given by

$$\tfrac{1}{2}\gamma_\theta = \tfrac{1}{2}(\omega_\theta - \omega_{\theta+90°})$$

$$= -\frac{\partial u}{\partial x}\cos\theta\sin\theta + \frac{\partial v}{\partial y}\cos\theta\sin\theta + \frac{1}{2}\left(\frac{\partial u}{\partial y} + \frac{\partial v}{\partial x}\right)(\cos^2\theta - \sin^2\theta)$$

$$(2.13)$$

Equations (2.8) and (2.13) enable normal and shear strains for any given value of the angle θ to be calculated if the terms $\partial u/\partial x$, $\partial v/\partial y$ and $\partial u/\partial y + \partial v/\partial x$ are known. The first two of these terms can be interpreted physically as the direct strains $e_\theta = e_{xx}$ and $e_\theta = e_{yy}$ in line segments lying in the x and y directions respectively. The third term is the shear strain $\gamma_\theta = \gamma_{xy}$ when OA and OB are aligned with the x and y directions, respectively. Hence the three *strain components*

$$e_{xx} = \frac{\partial u}{\partial x}, \qquad e_{yy} = \frac{\partial v}{\partial y}, \qquad \gamma_{xy} = \frac{\partial u}{\partial y} + \frac{\partial v}{\partial x} \qquad (2.14)$$

together define the *state of strain* at O.

In a general three-dimensional state of strain, it can be shown that the above analysis of deformations in the xy plane is still valid. The analysis must now, however, be extended to include deformations in the yz and zx planes, which leads to the definition of a total of *six* strain components to specify completely the state of strain at a point. These components are given in terms of the three displacements u, v and w as

$$e_{xx} = \frac{\partial u}{\partial x}, \qquad \gamma_{xy} = \frac{\partial u}{\partial y} + \frac{\partial v}{\partial x}$$

$$e_{yy} = \frac{\partial v}{\partial y}, \qquad \gamma_{yz} = \frac{\partial v}{\partial z} + \frac{\partial w}{\partial y} \qquad (2.15)$$

$$e_{zz} = \frac{\partial w}{\partial z}, \qquad \gamma_{zx} = \frac{\partial w}{\partial x} + \frac{\partial u}{\partial z}$$

For the special case of a *constant-strain state*, where the strain components are all independent of x, y and z, the corresponding solutions for the displacements obtained by integrating equations (2.15) are given by the linear functions

$$u = C_1 + C_2 x + C_3 y + C_4 z$$

$$v = D_1 + D_2 x + D_3 y + D_4 z \qquad (2.16)$$

$$w = E_1 + E_2 x + E_3 y + E_4 z$$

where the coefficients $C_1, D_1, E_1, \ldots,$ are all constants, and

$$e_{xx} = C_2, \qquad e_{yy} = D_3, \qquad e_{zz} = E_4$$

$$\gamma_{xy} = C_3 + D_2, \qquad \gamma_{yz} = D_4 + E_3, \qquad \gamma_{zx} = E_2 + C_4$$

Rigid-body displacements do not involve deformation and are obtained by combining (2.16) with the zero-strain conditions

$$C_2 = 0, D_3 = 0, E_4 = 0, D_2 = -C_3, E_2 = -C_4, E_3 = -D_4$$

to give

$$u = C_1 + C_3 y + C_4 z$$
$$v = D_1 - C_3 x + D_4 z \qquad\qquad (2.17)$$
$$w = E_1 - C_4 x - D_4 y$$

The six coefficients C_1, C_3, C_4, D_1, D_4 and E_1 in (2.17) are found by specifying the three components of translation of any point in the body, together with the rotations of the body about any three mutually orthogonal axes.

2.3 MOHR'S CIRCLE OF STRAIN

The strain–displacement equations (2.14) for plane strain, together with the trigonometrical identities for $\cos 2\theta$ and $\sin 2\theta$ in terms of $\cos \theta$ and $\sin \theta$ can be used to re-express (2.8) and (2.13) as

$$e_\theta = \tfrac{1}{2}(e_{xx} + e_{yy}) + \tfrac{1}{2}(e_{xx} - e_{yy}) \cos 2\theta + \tfrac{1}{2}\gamma_{xy} \sin 2\theta \qquad\qquad (2.18)$$

$$\tfrac{1}{2}\gamma_\theta = - \tfrac{1}{2}(e_{xx} - e_{yy}) \sin 2\theta + \tfrac{1}{2}\gamma_{xy} \cos 2\theta$$

Comparing equations (2.18) with (1.5), we observe that the former may be obtained from the latter by replacing each direct stress by the corresponding direct strain, and each shear stress by *one-half* of the corresponding shear strain. It must, therefore, be possible to draw a *Mohr's circle of strain*, with ordinates $(1/2)\gamma_\theta$ and abscissae e_θ (see Fig. 2.5), whose geometric properties are analogous to those of a Mohr's stress circle (see section 1.3.2). For a two-dimensional state of strain the following are true.

(1) There are two orthogonal directions, defined by the *principal axes of strain*, in which the direct strain e_θ has either a maximum or a minimum value. These values are the *principal strains*

$$e_{1,2} = \tfrac{1}{2}(e_{xx} + e_{yy}) \pm \tfrac{1}{2}\sqrt{(e_{xx} - e_{yy})^2 + \gamma_{xy}^2} \qquad\qquad (2.19)$$

The shear strain γ_θ measured with respect to the principal axes is equal to zero, and the orientation of these axes is given by the angle

$$2\beta = \tan^{-1}\left(\frac{\gamma_{xy}}{e_{xx} - e_{yy}}\right) \qquad\qquad (2.20)$$

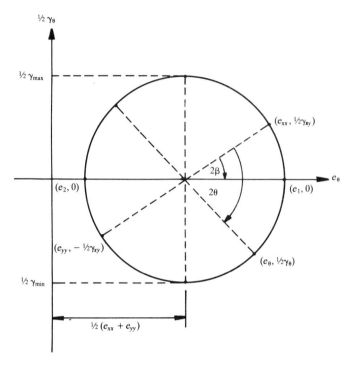

Fig. 2.5—Mohr's circle of strain.

(2) The shear strain γ_θ has the maximum value

$$\gamma_{max} = e_1 - e_2 \tag{2.21}$$

when measured with respect to the two orthogonal directions defined by the *axes of maximum shear strain*. These axes are at 45° to the principal axes of strain.

2.4   ROTATION OF AXES

In view of the analogy between two-dimensional states of stress and strain, we can deduce the transformation rule for a state of strain from the rule given by (1.11) for a state of stress. Thus

$$[e'] = [T]^{\mathrm{T}}[e][T] \tag{2.22}$$

where

$$[e] = \begin{bmatrix} e_{xx} & \tfrac{1}{2}\gamma_{xy} \\ \tfrac{1}{2}\gamma_{yx} & e_{yy} \end{bmatrix} \tag{2.23}$$

is the *strain tensor*, a *symmetric* matrix defining the state of strain in terms of x, y axes. $[e']$ is the corresponding strain tensor redefining the state of strain for x', y' axes obtained by an anti-clockwise rotation of the x, y axes through an angle θ. $[T]$

is the transformation matrix defined in (1.3) in terms of the direction cosines of the axes x', y'.

The analogy between states of stress and strain extends to three-dimensions where the state of strain is defined by

$$[e] = \begin{bmatrix} e_{xx} & \frac{1}{2}\gamma_{xy} & \frac{1}{2}\gamma_{xz} \\ \frac{1}{2}\gamma_{yx} & e_{yy} & \frac{1}{2}\gamma_{yz} \\ \frac{1}{2}\gamma_{zx} & \frac{1}{2}\gamma_{zy} & e_{zz} \end{bmatrix} \qquad (2.24)$$

The transformation rule given in (2.22) is applicable, provided that $[T]$ has the three-dimensional form defined in (1.20). For each property of a three-dimensional state of stress (see section 1.4), there is a corresponding property for a three-dimensional state of strain. It follows that there will be three principal strains, derived from a cubic equation analogous to (1.22), and three principal axes of strain. The maximum shear strain at any point is given by analogy with (1.23) as

$$\gamma_{max} = e_{max} - e_{min} \qquad (2.25)$$

where e_{max} and e_{min} are the maximum and minimum principal strains, respectively.

2.5 STRAIN COMPATIBILITY

In section 1.5, we saw that the variation in the stress state within a solid is governed by a set of differential equations. These were derived by requiring every portion of the solid to be in a state of equilibrium. We consider now the requirements which govern the variation in the state of strain within a solid.

The strain–displacement equations (2.15) are definitions for the six strain components in terms of the displacements. They can also be treated, however, as a set of six simultaneous equations in only the three unknown displacements u, v and w. The solution for these displacements will only be *unique* if the strains satisfy *compatibility relationships*, which are derived by eliminating u, v and w between the six equations. We can obtain the first of these relationships by differentiating γ_{xy} with respect to both x and y to give

$$\frac{\partial^2 \gamma_{xy}}{\partial x\, \partial y} = \frac{\partial^2}{\partial x\, \partial y}\left(\frac{\partial u}{\partial y}\right) + \frac{\partial^2}{\partial x\, \partial y}\left(\frac{\partial v}{\partial x}\right)$$

Remembering that u and v are assumed to be continuous functions of x and y, we can alter the order of differentiation on the right-hand side to give

$$\frac{\partial^2 \gamma_{xy}}{\partial x\, \partial y} = \frac{\partial^2}{\partial y^2}\left(\frac{\partial u}{\partial x}\right) + \frac{\partial^2}{\partial x^2}\left(\frac{\partial v}{\partial y}\right)$$

Finally, we can replace the displacement gradients on the right-hand side by the corresponding strains. The result, together with a further five relationships obtained in a similar way, are given by

$$\frac{\partial^2 e_{xx}}{\partial y^2} + \frac{\partial^2 e_{yy}}{\partial x^2} = \frac{\partial^2 \gamma_{xy}}{\partial x \, \partial y} \, , \qquad 2\frac{\partial^2 e_{xx}}{\partial y \, \partial z} = \frac{\partial}{\partial x}\left(-\frac{\partial \gamma_{yz}}{\partial x} + \frac{\partial \gamma_{zx}}{\partial y} + \frac{\partial \gamma_{xy}}{\partial z}\right)$$

$$\frac{\partial^2 e_{yy}}{\partial z^2} + \frac{\partial^2 e_{zz}}{\partial y^2} = \frac{\partial^2 \gamma_{yz}}{\partial y \, \partial z} \, , \qquad 2\frac{\partial^2 e_{yy}}{\partial z \, \partial x} = \frac{\partial}{\partial y}\left(\frac{\partial \gamma_{yz}}{\partial x} - \frac{\partial \gamma_{zx}}{\partial y} + \frac{\partial \gamma_{xy}}{\partial z}\right)$$

$$\frac{\partial^2 e_{zz}}{\partial x^2} + \frac{\partial^2 e_{xx}}{\partial z^2} = \frac{\partial^2 \gamma_{zx}}{\partial z \, \partial x} \, , \qquad 2\frac{\partial^2 e_{zz}}{\partial x \, \partial y} = \frac{\partial}{\partial z}\left(\frac{\partial \gamma_{yz}}{\partial x} + \frac{\partial \gamma_{zx}}{\partial y} - \frac{\partial \gamma_{xy}}{\partial z}\right)$$

$$(2.26)$$

Provided that strain compatibility is satisfied at every point in the interior of the solid the solutions for the displacements will be continuous and single-valued functions. On the surface of the solid, these solutions must satisfy any *displacement boundary conditions* prescribed there.

2.6   USE OF CYLINDRICAL COORDINATES

For future reference, we include here the equations for strain analysis using cylindrical coordinates (see section 1.6).

A three-dimensional state of strain is defined with respect to the r, θ, z directions by the strain tensor

$$[e'] = \begin{bmatrix} e_{rr} & \frac{1}{2}\gamma_{r\theta} & \frac{1}{2}\gamma_{rz} \\ \frac{1}{2}\gamma_{\theta r} & e_{\theta\theta} & \frac{1}{2}\gamma_{\theta z} \\ \frac{1}{2}\gamma_{zr} & \frac{1}{2}\gamma_{z\theta} & e_{zz} \end{bmatrix} \qquad (2.27)$$

which is related to $[e]$ by the transformation rule (2.22) where $[T]$ is given by (1.28).

The strain–displacement equations are

$$e_{rr} = \frac{\partial u}{\partial r} \, , \qquad\qquad \gamma_{r\theta} = \frac{1}{r}\frac{\partial u}{\partial \theta} - \frac{v}{r} + \frac{\partial v}{\partial r}$$

$$e_{\theta\theta} = \frac{u}{r} + \frac{1}{r}\frac{\partial v}{\partial \theta} \, , \qquad\qquad \gamma_{\theta z} = \frac{\partial v}{\partial z} + \frac{1}{r}\frac{\partial w}{\partial \theta} \qquad (2.28)$$

$$e_{zz} = \frac{\partial w}{\partial z} \, , \qquad\qquad \gamma_{zr} = \frac{\partial u}{\partial z} + \frac{\partial w}{\partial r}$$

where u, v and w are the components of displacement in the r, θ and z directions, respectively.

PROBLEMS

2.1   By integrating the strain–displacement equations for plane strain, show that the displacement components u and v of any point in a solid which undergoes rigid-body displacement in the xy plane are given by

$$u = ay + b$$

$$v = -ax + c$$

where a, b and c are constants.

2.2 Find the principal strains, the orientation of the principal axes of strain, and the maximum shear strain for the following cases of plane strain.

(i) $e_{xx} = 1200$, $e_{yy} = -400$, $\gamma_{xy} = -1200\mu$.
(ii) $e_{xx} = 1600$, $e_{yy} = -800$, $\gamma_{xy} = -1000\mu$.
(iii) $e_{rr} = 1000$, $e_{\theta\theta} = 800$, $\gamma_{r\theta} = 1000\mu$.

where the symbol μ represents micro-strain (10^{-6}).

(Answers:. (i) 1400μ, -600μ; $-18.4°$, $71.6°$; 2000μ.
(ii) 1700μ, -900μ; $-11.3°$, $78.7°$; 2600μ.
(iii) 1409μ, 390μ; $39.3°$, $129.3°$; 1409μ.)

2.3 Calculate the principal strains and maximum shear strain at a point in a solid where the state of strain is defined by

$$[e] = \begin{bmatrix} 30 & 50 & 80 \\ 50 & 10 & 0 \\ 80 & 0 & 20 \end{bmatrix} \mu$$

(Answers: 118.3μ, 12.9μ, -71.1μ ;$189.3\ \mu$.)

2.4 Electrical resistance strain gauges are used to measure the direct strains in the surface of a solid. Three such gauges measure e_θ in directions $\theta = 0°$, $45°$ and $90°$, where θ is measured anti-clockwise from the reference x direction. Given that $e_{0°} = 592\mu$, $e_{45°} = 308\mu$, $e_{90°} = -432\mu$ calculate the principal strains, and the orientation of the principal axes of strain.

(Answers: 641μ, -481μ; $12°$, $102°$.)

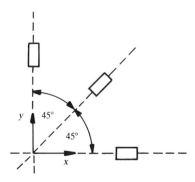

2.5   Direct strains in the surface of a solid are measured in each of three directions making angles $0°$, $60°$ and $120°$ with the reference x direction. Show that the principal strains are given by

$$e_{1,2} = \tfrac{1}{3}(e_{0°} + e_{60°} + e_{120°}) \pm \tfrac{1}{3}\sqrt{(2e_{0°} - e_{60°} - e_{120°})^2 + 3(e_{60°} - e_{120°})^2}$$

Chapter 3

Material Behaviour

3.1 INTRODUCTION

In previous chapters, we have seen how states of stress and strain at a point in a solid continuum are defined and analysed. Further, we have derived the equations which govern the variation in these states from one point to another. At no time, however, have we made any assumptions regarding the mechanical behaviour of the material, except that it can be adequately modelled as a continuum.

When we are dealing with a statically determinate problem, only the equations of stress equilibrium and their boundary conditions need be included in its formulation. In general, however, the problem of solving for the stresses is statically indeterminate and a description of the material behaviour must be included. This description takes the form of a set of *constitutive equations* relating the components of stress and strain.

3.2 CONSTITUTIVE EQUATIONS

To describe the stress–strain characteristics of an engineering material, we choose a model which approximates the actual behaviour with an adequate degree of accuracy.

An *elastic* material is defined as one in which the deformations caused by external loads disappear when these loads are removed (see Fig. 3.1). The stress–strain relationships in this model may be *linear*, in which case they are described by *Hooke's law*, or they may be *non-linear*. The behaviour of many materials is well represented by the linear elastic model, provided that the stresses are below a certain limit, termed the *yield point* of the material.

Above the yield point the material behaviour becomes *plastic* (see Fig. 3.2), and permanent strains remain when the external loading is removed.

For both elastic and plastic behaviour a material is assumed to respond

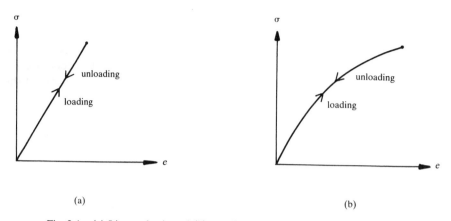

Fig. 3.1—(a) Linear elastic and (b) non-linear elastic stress–strain relationships.

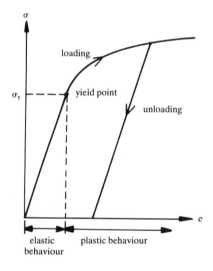

Fig. 3.2—Elastic–plastic constitutive relationship.

instantaneously to applied loads. When a material exhibits *creep* behaviour, in which the deformations caused by a steady load are functions of time, a *visceoelastic* or *viscoplastic* model is appropriate.

3.2.1 Generalised Hooke's law
From now on, we restrict our attention to *linear elastic* behaviour for which Hooke's law in its generalised form asserts that each component of strain is a linear function of the components of stress. For example,

$$e_{xx} = C_{11}\sigma_{xx} + C_{12}\sigma_{yy} + C_{13}\sigma_{zz} + C_{14}\tau_{xy} + C_{15}\tau_{yz} + C_{16}\tau_{zx}$$

where the coefficients $C_{11}, C_{12}, \ldots, C_{16}$ are the same for every point, provided that the material is *homogeneous*.

If the material properties are the same in all directions, the material is said to be *isotropic*. In this case, direct stresses give rise only to direct strains, shear stresses produce only shear strains and the stress–strain relationships are, therefore, of the form

$$
\begin{bmatrix} e_{xx} \\ e_{yy} \\ e_{zz} \\ \gamma_{xy} \\ \gamma_{yz} \\ \gamma_{zx} \end{bmatrix}
=
\begin{bmatrix}
C_{11} & C_{12} & C_{13} & 0 & 0 & 0 \\
C_{21} & C_{22} & C_{23} & 0 & 0 & 0 \\
C_{31} & C_{32} & C_{33} & 0 & 0 & 0 \\
0 & 0 & 0 & C_{44} & 0 & 0 \\
0 & 0 & 0 & 0 & C_{55} & 0 \\
0 & 0 & 0 & 0 & 0 & C_{66}
\end{bmatrix}
\begin{bmatrix} \sigma_{xx} \\ \sigma_{yy} \\ \sigma_{zz} \\ \tau_{xy} \\ \tau_{yz} \\ \tau_{zx} \end{bmatrix}
\tag{3.1}
$$

For a *uniaxial stress state* σ_{xx} the only nonzero strains are

$$
e_{xx} = \frac{\sigma_{xx}}{E}, \qquad e_{yy} = e_{zz} = -\nu \frac{\sigma_{xx}}{E}
$$

where E is the *elastic modulus* and ν *Poisson's ratio* of the material. Similar expressions can be written down for the strains associated with the uniaxial stress states σ_{yy} and σ_{zz}. For the state of *pure shear* in which τ_{xy} is the only nonzero stress, the corresponding strain is $\gamma_{xy} = \tau_{xy}/G$, where G is the *shear modulus* of the material, and similarly for states of pure shear defined by τ_{yz} and τ_{zx}. In this way, we can express the 12 coefficients in (3.1) in terms of only the *three* elastic constants E, ν and G and hence write the generalised Hooke's law as

$$
\begin{aligned}
e_{xx} &= \frac{1}{E}[\sigma_{xx} - \nu(\sigma_{yy} + \sigma_{zz})], & \gamma_{xy} &= \frac{1}{G}\tau_{xy} \\[2mm]
e_{yy} &= \frac{1}{E}[\sigma_{yy} - \nu(\sigma_{zz} + \sigma_{xx})], & \gamma_{yz} &= \frac{1}{G}\tau_{yz} \\[2mm]
e_{zz} &= \frac{1}{E}[\sigma_{zz} - \nu(\sigma_{xx} + \sigma_{yy})], & \gamma_{zx} &= \frac{1}{G}\tau_{zx}
\end{aligned}
\tag{3.2}
$$

The three elastic constants are not all independent of one another but can be shown to be related by

$$
E = 2G(1 + \nu)
\tag{3.3}
$$

It is often useful to have explicit expressions for the stresses in terms of the strains. These are obtained by solving equations (3.2) to give

$$\sigma_{xx} = \lambda\varepsilon + 2Ge_{xx}, \qquad \tau_{xy} = G\gamma_{xy}$$

$$\sigma_{yy} = \lambda\varepsilon + 2Ge_{yy}, \qquad \tau_{yz} = G\gamma_{yz} \qquad (3.4)$$

$$\sigma_{zz} = \lambda\varepsilon + 2Ge_{zz}, \qquad \tau_{zx} = G\gamma_{zx}$$

where

$$\varepsilon = e_{xx} + e_{yy} + e_{zz} = \frac{\partial u}{\partial x} + \frac{\partial v}{\partial y} + \frac{\partial w}{\partial z}$$

and

$$\lambda = \frac{\nu E}{(1 + \nu)(1 - 2\nu)}$$

3.2.2 Temperature effects

Engineering components frequently operate at elevated temperatures and with high temperature gradients. In the stress analysis of such problems, it is essential to incorporate the effects of temperature.

Strains are induced in a material not only by the presence of stresses but also by a change in temperature. A change ΔT in temperature at a point in an isotropic solid gives rise to the *thermal strains*

$$e_{xx} = e_{yy} = e_{zz} = \alpha\,\Delta T \qquad (3.5)$$

where α is the *coefficient of thermal expansion*. These strains must be added to the strains caused by stresses given in (3.2).

Temperature changes will also effect the values of the elastic constants. For most materials, this effect is negligible, except where the temperature changes are very large.

3.3 STRAIN ENERGY IN AN ELASTIC SOLID

When an elastic solid is deformed by externally applied forces, the *work done* by these forces is stored internally as *strain energy* which is completely recoverable when the solid is unloaded.

Consider an elastic bar of length L and uniform cross-section area A which is subjected to an axial force F (see Fig. 3.3a). The bar is stretched axially by an amount δ and the corresponding work done by the force F is given by

$$W = \int F\,d\delta \qquad (3.6)$$

which is the area under the load–deflection curve in Fig. 3.3(b). (This definition for W is, of course, valid irrespective of the mechanical behaviour of the material.) The work done is stored as the strain energy

$$U = W = \int F\,d\delta \qquad (3.7)$$

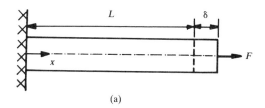

(a)

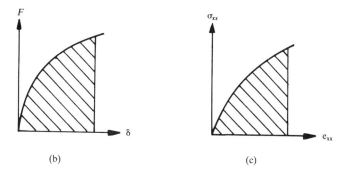

(b)   (c)

Fig. 3.3—(a) Geometry, (b) load–extension curve and (c) stress–strain curve for an axially
loaded elastic bar.

where F and δ are given in terms of the axial stress σ_{xx} and axial strain e_{xx} in the bar
as

$$F = A\sigma_{xx}, \qquad \delta = Le_{xx} \tag{3.8}$$

By substituting (3.8) into (3.7), we can write

$$U = AL \int \sigma_{xx}\, de_{xx} = \text{volume} \times \hat{U} \tag{3.9}$$

where $\hat{U}$, the strain energy per unit volume, or the *strain energy density*, is the area
under the stress–strain curve in Fig. 3.3(c).

For *linear* elastic behaviour, this curve becomes a straight line and therefore

$$\hat{U} = \int \sigma_{xx}\, de_{xx} = \tfrac{1}{2}\sigma_{xx}e_{xx} \tag{3.10}$$

Expressions for $\hat{U}$ similar to (3.10) can be obtained for the effects of the other five
stress components in a general stress state. The total strain energy of a linear elastic
solid is then given by the volume integral

$$U = \int \hat{U}\, dV \tag{3.11}$$

where

$$\hat{U} = \tfrac{1}{2}(\sigma_{xx}e_{xx} + \sigma_{yy}e_{yy} + \sigma_{zz}e_{zz} + \tau_{xy}\gamma_{xy} + \tau_{yz}\gamma_{yz} + \tau_{zx}\gamma_{zx})$$

3.4   YIELD CRITERIA

In engineering design, it is vital to identify the various possible *modes of failure* of a component under service conditions. *Excessive deformation* is one such mode of failure in which the dimensional changes of a component under load may be too great to allow the component to continue to fulfil its intended design function.

Although this may happen whilst the component is behaving elastically, it is more likely to be associated with the large strains involved in plastic deformation. It is often desirable, therefore, to ensure that the maximum stresses are never large enough to cause yielding. A *yield criterion* provides a means of identifying those critical stress states for which yielding is initiated.

Experimentally measured values of the *yield stress* σ_Y for the uniaxial stress state $[\sigma] = \sigma_1$ present in the tensile test are available for a wide range of engineering materials. The yield criterion here is simply

$$\sigma_1 = \sigma_Y \tag{3.12}$$

Many empirical criteria have been proposed which generalise (3.12) and enable yielding in a three-dimensional stress state to be correlated with yielding in the tensile test.

The simplest of these is the *Tresca* or *maximum shear stress criterion* which assumes that yielding occurs when the maximum shear stress τ_{max} reaches the value $(1/2)\sigma_Y$ which it attains in the tensile test at the yield point. With the aid of (1.23), we can write this condition as

$$\tau_{max} = \tfrac{1}{2}(\sigma_{max} - \sigma_{min}) = \tfrac{1}{2}\sigma_Y \tag{3.13}$$

A geometrical interpretation of (3.13) is given in Fig. 3.4 for plane stress ($\sigma_3 = 0$). Points whose coordinates give all possible states of plane stress which satisfy (3.13) provide a *yield locus* in the form of a *hexagon* in the $\sigma_1\sigma_2$ plane. Points inside the locus represent stress states where the material behaviour is elastic, whilst points outside correspond to plastic behaviour.

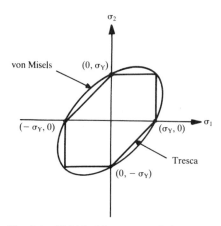

Fig. 3.4—Yield loci for a state of plane stress.

Any three-dimensional elastic stress state can be thought of as a *hydrostatic state* plus a *distortion state*, as shown in Fig. 3.5 where $\bar{\sigma}$ is the *hydrostatic* or *mean stress* given by

$$\bar{\sigma} = \tfrac{1}{3}(\sigma_1 + \sigma_2 + \sigma_3) \qquad (3.14)$$

and σ_1', σ_2' and σ_3' are the *deviatoric stresses* given by

$$\sigma_1' = \sigma_1 - \bar{\sigma}, \qquad \sigma_2' = \sigma_2 - \bar{\sigma}, \qquad \sigma_3' = \sigma_3 - \bar{\sigma} \qquad (3.15)$$

The deformations caused by the hydrostatic and distortion states produce changes in the volume and the shape of the solid, respectively. The *von Mises* or *maximum distortion energy criterion* assumes that yielding occurs when the strain energy density $\hat{U}_D$ of the distortion state reaches the value which it attains in the tensile test at the yield point. From (3.11), we obtain

$$\hat{U}_D = \tfrac{1}{2}(\sigma_1'e_1' + \sigma_2'e_2' + \sigma_3'e_3') \qquad (3.16)$$

where the strains e_1', e_2' and e_3' are given by (3.2) as

$$e_1' = \frac{1}{E}\left[\sigma_1' - \nu(\sigma_2' + \sigma_3')\right], \qquad \text{etc.} \qquad (3.17)$$

After substituting (3.17) into (3.16) and then eliminating the deviatoric stresses using (3.15), we conclude that

$$\hat{U}_D = \frac{1+\nu}{6E}\left[(\sigma_1 - \sigma_2)^2 + (\sigma_2 - \sigma_3)^2 + (\sigma_3 - \sigma_1)^2\right] \qquad (3.18)$$

At the yield point in the tensile test we have $\sigma_1 = \sigma_Y$, and $\sigma_2 = \sigma_3 = 0$ for which (3.18) becomes

$$\hat{U}_D = \frac{1+\nu}{3E}\,\sigma_Y^2 \qquad (3.19)$$

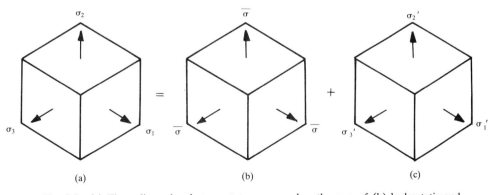

Fig. 3.5—(a) Three-dimensional stress state, expressed as the sum of (b) hydrostatic and (c) distortion stress states.

By equating (3.18) and (3.19), we obtain the von Mises yield criterion for a three-dimensional stress state as

$$\tfrac{1}{2}[(\sigma_1 - \sigma_2)^2 + (\sigma_2 - \sigma_3)^2 + (\sigma_3 - \sigma_1)^2] = \sigma_Y^2 \tag{3.20}$$

The geometric interpretation of (3.20) for plane stress is shown in Fig. 3.4 where the yield locus is seen to be given by an *ellipse* in the $\sigma_1 \sigma_2$ plane. Note that the Tresca and von Mises criteria coincide only for uniaxial stress states and that the latter is generally the less conservative criterion.

For design purposes the yield stress σ_Y used in the yield criteria (3.13) and (3.20) is normally replaced by σ_Y/n where $n(>1)$ is referred to as a *factor of safety* against yielding. This factor is used to account for the uncertainties inherent in the determination of σ_Y and in the analysis of the stresses in the component.

3.5 FRACTURE CRITERIA

Of all possible modes of component failure, *fracture* is usually the most serious. Even though the stresses in a component may be generally below the yield conditon, the inevitable presence of *cracks* will give rise to local intensifications. The stresses adjacent to one of these cracks can be sufficiently high to cause it to extend in an uncontrolled manner, thereby leading to the catastrophic failure of the component. A *fracture criterion* provides a means of predicting the loading on a component which will cause existing cracks to extend or, conversely, for a set of prescribed loads it will enable the critical size of a crack at a particular location to be calculated.

The stresses adjacent to a crack tip will usually be high enough to cause yielding and the development of a *plastic zone*. Provided that only *small-scale yielding* occurs in the immediate vicinity of the tip, we can use fracture parameters to characterise the onset of crack extension which are derived from a linear elastic stress analysis. These parameters form the basis of *linear elastic fracture mechanics* (LEFM) which is becoming widely accepted in engineering practice as a valuable aid in the estimation of safe design loads.

It was Griffith in 1920 who was the first person to note that a necessary condition for a crack to extend is that the process involved must be *energetically* possible. The rate at which the potential energy Π of a body containing a crack of surface area A decreases as the crack extends is given by

$$\mathcal{G}_1 = -\frac{d\Pi}{dA} = -\frac{d}{dA}(U - W) \tag{3.21}$$

where U is the strain energy in the body, and $W = 2U$ is the work done by the external loads F. Provided that the loads remain constant during crack extension, it can be shown that

$$\mathcal{G}_1 = \left(\frac{\partial U}{\partial A}\right)_F = \frac{1}{2}\left(\frac{\partial W}{\partial A}\right)_F \tag{3.22}$$

Denoting the rate at which energy is dissipated in the material during the fracture process as dD/dA, we can, according to Griffith, write the fracture criterion as

$$\mathcal{G}_I = \frac{dD}{dA} = \mathcal{G}_{Ic} \tag{3.23}$$

The critical value $\mathcal{G}_{Ic}$ is identified as the *rate at which plastic work is done* during crack extension and must be measured experimentally.

In general, there are three possible *modes* of crack extension (see Fig. 3.6) where $\mathcal{G}_I$, $\mathcal{G}_{II}$ and $\mathcal{G}_{III}$ denote the potential energy release rates in each mode. An example of a mode I problem for an infinite plate of unit thickness, containing a crack of length $2a$, and subjected to a uniaxial tensile stress σ, is shown in Fig. 3.7.

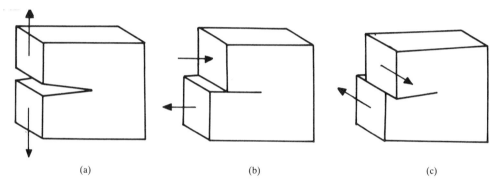

(a)   (b)   (c)

Fig. 3.6—Modes of crack extension: (a) opening mode I; (b) sliding mode II; (c) tearing mode III.

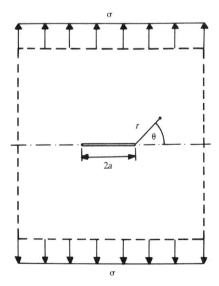

Fig. 3.7—Cracked infinite plate subjected to a uniaxial stress.

For plane stress conditions the strain energy in the plate is $U = \pi\sigma^2 a^2/E$ and thus (3.22) becomes

$$\mathcal{G}_1 = \frac{1}{2}\left(\frac{\partial U}{\partial a}\right)_\sigma = \frac{\pi\sigma^2 a}{E}. \tag{3.24}$$

In the case of plane strain, E is replaced by $E^* = E/(1 - \nu_2)$. The *fracture stress* $\sigma = \sigma_F$ is found by equating (3.24) and (3.23) to give

$$\sigma_F = \sqrt{\frac{E\,\mathcal{G}_{\mathrm{Ic}}}{\pi a}} \tag{3.25}$$

An equivalent approach to fracture in LEFM involves an analysis of stresses in the near vicinity of the crack tip rather than an energy balance for the whole body. The elastic stresses local to the crack tip for a mode I problem (see Fig. 3.7) will be shown in a later chapter to be of the form

$$\sigma_{ij} = \frac{K_{\mathrm{I}}}{\sqrt{r}} f_{ij}(\theta) \tag{3.26}$$

where K_{I}, the *stress intensity factor*, is a single parameter characterising the stress field close to the crack tip where the $1/\sqrt{r}$ *singularity* dominates. In practice, of course, yielding occurs in this region to relieve the theoretically infinite stresses. Nonetheless, for small-scale yielding, it is realistic to view the plastic zone as being embedded within an elastic field characterised by K_{I}. It is then postulated that fracture occurs when a critical value

$$K_{\mathrm{I}} = K_{\mathrm{Ic}} \tag{3.27}$$

is reached. K_{Ic} is the *fracture toughness* of the material and, like $\mathcal{G}_{\mathrm{Ic}}$, must be measured experimentally.

For the crack problem in Fig. 3.7, it can be shown that

$$K_{\mathrm{I}} = \sigma\sqrt{\pi a} \tag{3.28}$$

where the fracture stress σ_F is found by equating (3.27) and (3.28) to give

$$\sigma_F = \frac{K_{\mathrm{Ic}}}{\sqrt{\pi a}} \tag{3.29}$$

The above approaches were shown to be equivalent by Irwin who, by calculating the energy required to close up a small portion of a crack, derived a relationship between the fracture parameters which, for plane conditions, is given by

$$\mathcal{G}_1 = \frac{K_{\mathrm{I}}^2}{E} \tag{3.30}$$

It is easily confirmed, using this relationship, that the fracture stresses, predicted by (3.25) and (3.29) for the two approaches, are equal to one another.

The fracture criteria of LEFM are based on the assumption that the size of the crack tip plastic zone is small in relation to the crack length. Otherwise the problem is one of *ductile fracture* where $\mathscr{G}_1$ and K_1 are no longer the characterising parameters. Various attempts have been made to extend the validity of LEFM by, for example, using an *equivalent crack length* in the formula for K_1 to take account of the presence of the plastic zone. According to Irwin, equation (3.28) should be replaced by

$$K_1 = \sigma\sqrt{\pi(a + r_Y)} \tag{3.31}$$

where

$$r_Y = \frac{1}{2\pi}\left(\frac{K_1}{\sigma_Y}\right)^2$$

is a measure of the extent of the plastic zone.

One of the commonest causes of mechanical failure, in which fracture plays a key role, is *fatigue*. This is a process in which the repeated application of loading to a component causes cracks to propagate. At first, crack propagation is stable and may continue to be so for many thousands, or even millions, of load cycles. Finally, however, the process becomes unstable and leads to catastrophic failure of the component. Given that the stress intensity factor characterises the crack tip stress environment for small-scale yielding, it might be expected to determine the rate da/dn of fatigue crack propagation per load cycle. For example, in the case of loads which vary between zero and some positive value, the following empirical formula is used:

$$\frac{da}{dn} = C(\Delta K_1)_m \tag{3.32}$$

Here ΔK_1 denotes the range of the stress intensity factor during a load cycle, and C and m are constants for a particular material which are obtained by experiment. Using such formulae, it is possible to estimate the residual fatigue life of a component with cracks of a known size.

PROBLEMS

3.1   An elastic thin-walled cylindrical tank of mean diameter d and wall thickness t has closed ends and is subjected to an internal pressure p. Show that the circumferential stress in the wall is twice the longitudinal stress and that the change ΔV in the internal volume V of the tank due to the pressure p is given by

$$\frac{\Delta V}{V} = \frac{pd}{4Et}(5 - 4v)$$

3.2   The state of stress at a point in a linear elastic solid is defined by the principal stresses $\sigma_1, \sigma_2, \sigma_3$. By considering the change ΔV in volume of a small cube of the solid of initial volume V show that

$$\frac{\Delta V}{V} = \frac{1 - 2v}{E}(\sigma_1 + \sigma_2 + \sigma_3)$$

3.3   Three strain gauges are used to measure the direct strains e_θ in the surface of a plate in directions $\theta = 0°$, $60°$ and $120°$, where θ is measured anti-clockwise from the reference x direction. If $e_{0°} = 554\mu$, $e_{60°} = -456\mu$ and $e_{120°} = 64\mu$, calculate the corresponding principal stresses, the orientations of the principal axes, and the principal strain in the direction normal to the surface of the plate. Assume that $E = 200$ GN/m^2 and $v = 0.3$.

(Answers: 105 MN/m^2, -74 MN/m^2; $-15.5°$, $74.5°$; 46μ.)

3.4   A state of pure shear in a linear elastic solid is defined by $\tau_{xy} = \tau$. By relating the principal stresses and strains using generalised Hooke's law, verify the relationship between the three elastic constants given by

$$E = 2G(1 + v)$$

3.5   The stress state in a thin-walled tube subjected to combined bending and torsion is defined by $\sigma_{xx} = \sigma$, $\tau_{xy} = \tau$. The yield stress of the material in uniaxial tension is σ_Y. Show that the yield criterion can be expressed in the form

$$\left(\frac{\sigma}{\sigma_Y}\right)^2 + \lambda\left(\frac{\tau}{\sigma_Y}\right)^2 = 1$$

where the value of the constant λ is equal to 4 for Tresca's yield criterion and 3 for von Mises' criterion.

Chapter 4

Formulation of Stress Analysis Problems

4.1 INTRODUCTION

In the first three chapters of this book, we have established the basis of the three-dimensional theory of elasticity. We have discussed the concepts of a state of stress and strain at a point and have derived the conditions which govern the variation in these states from one point to another in a solid. The relationships between stress and strain for a linear elastic material have also been presented.

We begin this chapter with a review of the general theory of elasticity. A number of simplified engineering theories, to be examined more fully in later chapters, are then outlined. Finally we include a selection of topics, a knowledge of which is useful in the formulation of practical problems.

4.2 THREE-DIMENSIONAL THEORY OF ELASTICITY

The conditions which must be satisfied by the solution of a problem in the three-dimensional theory of elasticity are as follows.

(1) Equilibrium: the six stress components must satisfy the three equations of equilibrium (1.25) in the interior, together with the stress boundary conditions (1.26) prescribed on the surface of the solid.
(2) Compatibility: the six components of strain must satisfy the six equations (2.26) of strain compatibility in the interior, in order that the displacements shall be continuous and single-valued functions. The displacements must also satisfy any displacement boundary conditions prescribed on the surface.
(3) Constitutive equations: the strains must be related to the stresses by way of the constitutive equations (3.2).

If an analytical solution of these equations is to be attempted it is convenient to first express them in one of the two following forms.

(1) Equilibrium in terms of displacements: first the equations of equilibrium (1.25) are expressed in terms of strains using the constitutive equations (3.4). By expressing the strains in terms of the displacements, using the strain–displacement equations (2.15), we are then able to write the three equilibrium equations in terms of the three displacements as

$$(\lambda + G)\frac{\partial \varepsilon}{\partial x} + G\nabla^2 u + b_x = 0$$

$$(\lambda + G)\frac{\partial \varepsilon}{\partial y} + G\nabla^2 v + b_y = 0 \qquad\qquad (4.1)$$

$$(\lambda + G)\frac{\partial \varepsilon}{\partial z} + G\nabla^2 w + b_z = 0$$

Since the problem is formulated entirely in terms of displacements, we do not need to satisfy explicitly the compatibility equations.

(2) Compatibility in terms of stresses: the six equations (2.26) of strain compatibility are expressed in terms of the six stress components using the constitutive equations (3.2). To simplify the task of solving these equations simultaneously with the equilibrium equations (1.25), a mathematical device is used. This involves defining the stresses in terms of *stress functions* Φ_1, Φ_2 and Φ_3 in such a way as to satisfy equilibrium automatically. In the absence of body forces, these definitions are

$$\sigma_{xx} = \frac{\partial^2 \Phi_3}{\partial y^2} + \frac{\partial^2 \Phi_2}{\partial z^2}, \qquad \tau_{xy} = \frac{-\partial^2 \Phi_3}{\partial x\, \partial y}$$

$$\sigma_{yy} = \frac{\partial^2 \Phi_1}{\partial z^2} + \frac{\partial^2 \Phi_3}{\partial x^2}, \qquad \tau_{yz} = \frac{-\partial^2 \Phi_1}{\partial y\, \partial z} \qquad\qquad (4.2)$$

$$\sigma_{zz} = \frac{\partial^2 \Phi_2}{\partial x^2} + \frac{\partial^2 \Phi_1}{\partial y^2}, \qquad \tau_{zx} = \frac{-\partial^2 \Phi_2}{\partial z\, \partial x}$$

The compatibility equations, together with the boundary conditions of the problem, are then expressed in terms of these functions.

4.3 ENGINEERING THEORIES OF ELASTICITY

Analytical solutions of problems in the three-dimensional theory of elasticity are only possible in a very limited number of cases. In engineering, however, we can frequently employ simplified two- and even one-dimensional theories which are adequate for the analysis of stresses in components with particular types of geometry and loading configuration. Each of these theories involves assumptions about the deformation, on the basis of which the solution for the equilibrium stress state is then sought. The most commonly used theories include the following.

(1) Plane theory: this deals with solids bounded by two plane surfaces and subjected to in-plane loading on the remaining boundary. Two distinct plane problems are identified in which either the state of stress or the state of strain is assumed to be two-dimensional (see Chapter 9).

(2) Axisymmetric theory: this is a two-dimensional theory for the stress analysis of bodies of revolution subjected to axisymmetric loading (see Chapter 10).

(3) Torsion theories: there are two distinct torsional problems which can be described by a two-dimensional theory. The first of these involves a shaft of arbitrary but uniform cross-section in which the state of strain is two-dimensional. (In the special case of a circular shaft the theory reduces to the one-dimensional elementary theory of torsion.) The second involves a circular shaft of variable diameter in which the plane cross-section assumption made in the elementary theory remains valid but stress variations in the axial direction are now significant (see Chapter 11).

(4) Plate and shell theories: these deal with solids bounded by surfaces whose dimensions are large compared with the distance between them. Loading is in the form of lateral forces and the solid is referred to as a *plate* or *shell* depending on whether the surfaces are initially plane or curved, respectively. In the case of a *thin* plate or shell the effect of shear forces on the deformation is assumed to be negligible.

4.4 ST VENANT'S PRINCIPLE

In the application of the various engineering theories of elasticity the principle of St Venant is frequently invoked. The principle states that *statically equivalent* systems of loading applied to a small area of the surface of a structure each produce identical elastic responses, except in the immediate vicinity of this area.

For two systems of loading to be statically equivalent, they must have the same force and moment resultants. If t_x, t_y, t_z and t'_x, t'_y, t'_z are the boundary tractions over a small area ΔA (see Fig. 4.1) for two such systems, then

$$\int_{\Delta A} t_x \, dA = \int_{\Delta A} t'_x \, dA, \qquad \int_{\Delta A} t_y \, dA = \int_{\Delta A} t'_y \, dA, \qquad \int_{\Delta A} t_z \, dA = \int_{\Delta A} t'_z \, dA$$

$$\int_{\Delta A} (xt_y - yt_x) \, dA = \int_{\Delta A} (xt'_y - yt'_x) \, dA$$

$$\int_{\Delta A} (yt_z - zt_y) \, dA = \int_{\Delta A} (yt'_z - zt'_y) \, dA \qquad (4.3)$$

$$\int_{\Delta A} (zt_x - xt_z) \, dA = \int_{\Delta A} (zt'_x - xt'_z) \, dA$$

Suppose, for example, that we have obtained a solution to the equations of elasticity which satisfies the stress boundary conditions for a problem, except over a

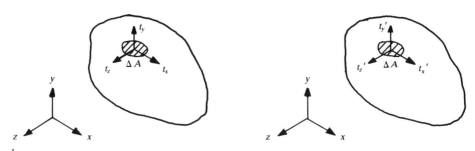

Fig. 4.1—Statically equivalent systems of loading.

small portion of the surface of the solid where there are some minor discrepancies. According to St Venant's principle, the solution will be valid except near this portion of the surface.

4.5 PRINCIPLE OF SUPERPOSITION

Let $\sigma_{xx}, \sigma_{yy}, \ldots, \tau_{zx}$ be the stresses in a linear elastic solid produced by the boundary tractions $\bar{t}_x, \bar{t}_y, \bar{t}_z$ and the body forces b_x, b_y, b_z. These stresses must satisfy the equations of equilibrium (1.25), the first of which requires that

$$\frac{\partial \sigma_{xx}}{\partial x} + \frac{\partial \tau_{xy}}{\partial y} + \frac{\partial \tau_{xz}}{\partial z} + b_x = 0 \tag{4.4}$$

and the boundary conditions (1.26), typified by

$$\bar{t}_x = l\sigma_{xx} + m\tau_{xy} + n\tau_{xz} \tag{4.5}$$

If tractions $\bar{t}'_x, \bar{t}'_y, \bar{t}'_z$ and body forces b'_x, b'_y, b'_z act instead, the stresses σ'_{xx}, $\sigma'_{yy}, \ldots, \tau'_{zx}$ are produced and equations (4.4) and (4.5) are replaced by

$$\frac{\partial \sigma'_{xx}}{\partial x} + \frac{\partial \tau'_{xy}}{\partial y} + \frac{\partial \tau'_{xz}}{\partial z} + b'_x = 0 \tag{4.6}$$

$$\bar{t}'_x = l\sigma'_{xx} + m\tau'_{xy} + n\tau'_{xz} \tag{4.7}$$

By adding (4.4) and (4.6), we obtain

$$\frac{\partial}{\partial x}(\sigma_{xx} + \sigma'_{xx}) + \frac{\partial}{\partial y}(\tau_{xy} + \tau'_{xy}) + \frac{\partial}{\partial z}(\tau_{xz} + \tau'_{xz}) + (b_x + b'_x) = 0$$

and, by adding (4.5) and (4.7), we obtain

$$\bar{t}_x + \bar{t}'_x = l(\sigma_{xx} + \sigma'_{xx}) + m(\tau_{xy} + \tau'_{xy}) + n(\tau_{xz} + \tau'_{xz})$$

Similar results are obtained by combining the other equilibrium equations and boundary conditions for the two systems of loading.

The conclusion is that stresses $\sigma_{xx} + \sigma'_{xx}, \ldots, \tau_{zx} + \tau'_{zx}$ satisfy all the conditions determining the stresses due to the surface tractions $\bar{t}_x + \bar{t}'_x, \bar{t}_y + \bar{t}'_y, \bar{t}_z + \bar{t}'_z$ and the

Fig. 4.2—Example of the principle of superposition.

body forces $b_x + b'_x$, $b_y + b'_y$, $b_z + b'_z$. This confirms the *principle of superposition* which states that the combined effect of two sets of external loads on a structure is found by summing the individual effects of each set (see Fig. 4.2). The principle holds because the differential equations and boundary conditions in the theory of elasticity are all *linear*.

4.6 STRUCTURAL SYMMETRY

Engineering structures frequently have one or more planes of *geometric symmetry*. The analysis of such problems can be simplified by considering only a symmetric portion of the complete structure bounded by the planes of symmetry. The boundary conditions to be imposed on such planes will depend on the distribution of external loads.

The simply supported beam AB shown in Fig. 4.3(a) carries external loading which is *symmetric* with respect to the plane of geometric symmetry $x = 0$. The components of displacement at the points $P(x, y, z)$ and $P'(-x, y, z)$ are u, v, w and u', v', w', respectively. By setting up a mirror parallel to the plane of symmetry we

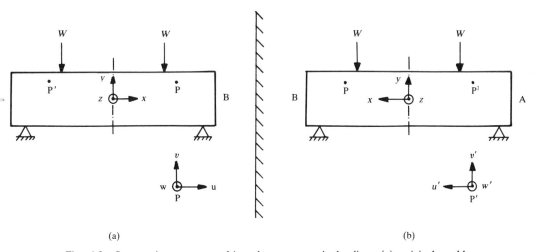

(a)   (b)

Fig. 4.3—Symmetric structure subjected to symmetric loading: (a) original problem; (b) image problem.

obtain an identical *image problem* shown in Fig. 4.3(b). We therefore conclude, after taking into account the reversal in direction of the x axis caused by the reflection, that the components of displacement u, v, w of the point P in the original problem must be the same as the components of displacement $-u', v', w'$ of the equivalent point P' in the image problem. Hence, in the original problem, u is an *odd function*, whilst v and w are both *even functions* of x. Thus, on the plane of symmetry $x = 0$, we have

$$u = 0 \qquad \text{and} \qquad \frac{\partial v}{\partial x} = \frac{\partial w}{\partial x} = 0 \qquad \text{for all } y \text{ and } z \qquad (4.8)$$

shear strains which are given by (2.15) as

$$\gamma_{xy} = \frac{\partial u}{\partial y} + \frac{\partial v}{\partial x} = 0 \qquad \text{and} \qquad \gamma_{zx} = \frac{\partial w}{\partial x} + \frac{\partial u}{\partial z} = 0$$

and, according to (3.4), shear stresses $\tau_{xy} = \tau_{xz} = 0$. The boundary conditions on the plane of symmetry when it bounds a symmetric half of the structure may be summarised as

$$u = 0 \qquad \text{and} \qquad \tau_{xy} = \tau_{xz} = 0 \qquad (4.9)$$

In other words, *the plane carries no shear stresses and there are no displacements normal to the plane.*

Geometric symmetry can also be exploited when the loading is *anti-symmetric*. Consider the cantilever beam shown in Fig. 4.4(a) which carries a uniform shear

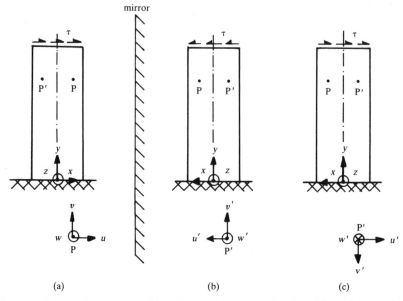

(a)   (b)   (c)

Fig. 4.4—Symmetric structure subjected to anti-symmetric loading: (a) original problem; (b) image problem; (c) image problem with loading reversed.

stress τ on its free end which is an anti-symmetric loading with respect to the plane of geometric symmetry $x = 0$. The image problem, obtained in the same manner as in the previous example, is shown in Fig. 4.4(b). When we reverse the sign of the loading, the signs of the displacements are also reversed and we obtain a final problem (see Fig. 4.4(c)) which is identical with the original one. It follows that the components of displacement u, v, w of the point P in the original problem are the same as the components u', $-v'$, $-w'$ of the equivalent point P$'$ in the final problem. Thus, in the original problem, u is an *even function*, and both v and w are *odd functions* of x. This implies that on the plane of symmetry $x = 0$ we have

$$\frac{\partial u}{\partial x} = 0 \qquad \text{and} \qquad v = w = 0 \qquad \text{for all } y \text{ and } z \qquad (4.10)$$

According to (2.15), the corresponding direct strains are

$$e_{xx} = \frac{\partial u}{\partial x} = 0, \qquad e_{yy} = \frac{\partial v}{\partial y} = 0, \qquad e_{zz} = \frac{\partial w}{\partial z} = 0$$

and, using (3.4), we deduce that $\sigma_{xx} = 0$. The conditions which must be prescribed on the plane of symmetry when it forms a boundary for a symmetric half of the problem can be summarised as

$$v = w = 0 \qquad \text{and} \qquad \sigma_{xx} = 0 \qquad (4.11)$$

In other words, *the plane carries no direct stress and there are no displacements in the plane itself.*

The analysis of *any* linear elastic structure with a plane of geometric symmetry can be simplified, regardless of the distribution of external loading, by first representing this loading as the sum of symmetric and anti-symmetric components. According to the principle of superposition (see section 4.5), the combined effect of the two components of loading is found by summing the effects of each component acting alone. An example of this is shown in Fig. 4.5 for a ring, rigidly constrained arounds its periphery, and carrying a radial load W on its inner surface.

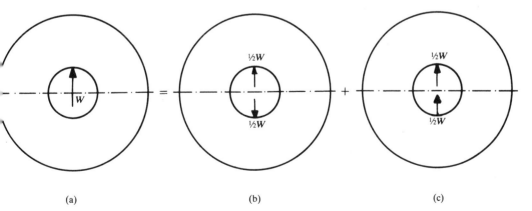

(a)   (b)   (c)

Fig. 4.5—An example of (a) unsymmetric loading, represented as the sum of (b) symmetric and (c) anti-symmetric loading.

4.7 DIMENSIONAL ANALYSIS

When an analytical solution to a stress analysis problem can be obtained, the formulae for the stresses and displacements must be *dimensionally correct* if they are to give consistent results regardless of the system of units employed. This requirement, which means that the formulae must possess certain features, forms the basis of the theory of *dimensional analysis* which is employed in various branches of engineering.

For problems which do not have an analytical solution, we need only to assume the existence of such formulae. Dimensional analysis can then be employed to provide useful information about the way in which the various quantities involved affect the solution. This information is particularly valuable in the interpretation of experimental measurements and of the results obtained using numerical solution techniques.

Consider an elastic structure subjected to a number of loads which are expressed in terms of ratios to a typical load, say P. Similarly the dimensions of the structure are expressed in terms of ratios to a typical dimension, say L. For a material with an elastic modulus E and Poisson's ratio v, the formula for a stress component σ at any point with coordinates (x, y, z) will be a relationship between the quantities

$$\sigma, x, y, z, E, v, P, L$$

The measurement of these quantities involves only the two *fundamental units* of force and length. According to *Buckingham's theorem*, these quantities can be combined into *dimensionless groups*, the number of independent groups being equal to the number of quantities minus the number of fundamental units. In the present problem there are the six groups

$$\frac{\sigma L^2}{P}, \frac{x}{L}, \frac{y}{L}, \frac{z}{L}, v, \frac{P}{EL^2}$$

and any other groups, such as σ/E, will be merely products of powers of these, and so not independent. A dimensionless group has the same numerical value, regardless of the system of units used for its evaluation.

The formula for the stress σ as a function of the other quantities must be a relationship between these groups of the form

$$\frac{\sigma L^2}{P} = f\left(\frac{x}{L}, \frac{y}{L}, \frac{z}{L}, v, \frac{P}{EL^2}\right)$$

where the right-hand side is a function which exists even when it cannot be determined analytically. In the case of a linear elastic material the stress σ is proportional to P. It follows that the group P/EL^2 must not appear in f and therefore that

$$\frac{\sigma L^2}{P} = f\left(\frac{x}{L}, \frac{y}{L}, \frac{z}{L}, v\right) \tag{4.12}$$

from which we conclude that the stress is *independent* of E. This conclusion holds for any linear elastic structure.

In a similar way the formula for the displacement u at any point (x, y, z) in the structure can be written as a relationship between dimensionless groups of the form

$$\frac{uEL}{P} = g\left(\frac{x}{L}, \frac{y}{L}, \frac{z}{L}, v\right) \tag{4.13}$$

When external loading other than point loads is present, the dimensionless groups must be modified accordingly as shown in Table 4.1.

Table 4.1 Dimensionless groups for various types of external loading

External loading	Dimensionless groups	
	Stress	Displacement
Point loads P	$\sigma L^2/P$	uEL/P
Surface stress p	σ/p	uE/pL
Weight force $\rho g L^3$	$\sigma/\rho g L$	$uE/\rho g L^2$
Centrifugal force $\rho \omega^2 L^4$	$\sigma/\rho \omega^2 L^2$	$uE/\rho \omega^2 L^3$
Torque T	$\tau L^3/T$	uGL^2/T
Bending moment M	$\sigma L^3/M$	uEL^2/M

PROBLEMS

4.1 Starting from the differential equations of equilibrium show that, in the absence of body forces, the displacements u and v in a plane stress problem must satisfy

$$\frac{\partial^2 u}{\partial x^2} + \frac{\partial^2 u}{\partial y^2} + \frac{1+v}{1-v} \frac{\partial}{\partial x}\left(\frac{\partial v}{\partial x} + \frac{\partial v}{\partial y}\right) = 0$$

and

$$\frac{\partial^2 v}{\partial x^2} + \frac{\partial^2 v}{\partial y^2} + \frac{1+v}{1-v} \frac{\partial}{\partial y}\left(\frac{\partial u}{\partial x} + \frac{\partial v}{\partial y}\right) = 0$$

assuming that the material is linearly elastic.

4.2 Verify that for a state of plane strain in a linearly elastic material the equation of compatibility can be expressed in terms of the stresses as

$$\left(\frac{\partial^2}{\partial x^2} + \frac{\partial^2}{\partial y^2}\right)(\sigma_{xx} + \sigma_{yy}) = 0$$

4.3 The solution for the stresses in problem 1.5 was obtained by satisfying the boundary conditions on the upper and lower faces of the beam. Use St Venant's principle to examine any possible discrepancies in the solution for the stresses on the end faces of the beam $x = 0$ and $x = L$.

4.4 Explain how the loading on the cantilever beam in problem 1.5 can be represented as the sum of components which are symmetric and anti-symmetric with respect to the plane of geometric symmetry $y = 0$.

4.5 A linear elastic structure is loaded by a point load P and contains a line crack of length a for which the stress intensity factor is K_1. Using dimensional analysis, derive the functional form of the formula relating K_1, P, a, E and v in terms of dimensionless groups. Hence show that the value of K_1 is independent of the elastic modulus E.

(Answer: $K_1/Ea^{1/2} = f(P/Ea^2, v)$.)

Chapter 5

Finite Element Concepts

5.1 INTRODUCTION

In the last chapter, we dealt with the mathematical formulation of stress analysis problems. This involves the differential equations of equilibrium and compatibility, together with the stress–strain relationships and the boundary conditions. Analytical solutions to these equations are seldom possible, and it is often necessary, therefore, to employ a *numerical method*.

A number of numerical stress analysis techniques are currently available, and their implementation is being greatly facilitated by the increasingly widespread availability of computers. The essential common feature of these methods is that the original problem, posed in terms of differential equations in the unknown *continuous* functions, is replaced by a formulation involving a set of algebraic equations in the *discrete* values of the unknowns at a finite number of points in the solid. In other words, the continuum model of the problem is approximated by a discrete model having a finite number of *degrees of freedom*.

Of the numerical methods available the *finite element method*, first developed in the 1950s for aircraft design, is the most widely used. The basis of the method is the representation of a structure by an assemblage of subdivisions or *finite elements*, as shown in Fig. 5.1. These elements are considered to be interconnected at joints, called *nodes* or *nodal points*, at which the values of the unknowns (usually the displacements) are to be approximated. If successively finer *discretisations* of the structure provide solutions which converge to the exact solution, then it is likely that a moderately coarse subdivision will provide a solution of acceptable accuracy. The computational effort required to obtain a solution will depend upon the number of degrees of freedom in the finite element model. In engineering practice a limit will be imposed on the degree of subdivision of the structure by the need to strike a balance between computing costs and solution accuracy.

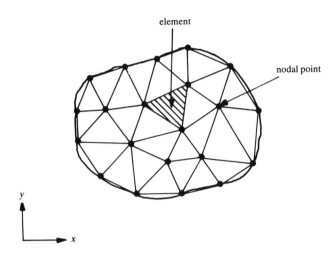

Fig. 5.1—Representation of a two-dimensional solid as an assemblage of triangular finite elements.

5.2 LINEAR SPRING ELEMENT

To illustrate the essential steps in finite element analysis, we begin by considering the elastic spring problem shown in Fig. 5.2. For each spring, there is assumed to be a linear relationship between the spring force P and the increase Δ in length given by

$$P = s\Delta \tag{5.1}$$

where s is the *spring stiffness*. Two springs AB and BC having stiffnesses s_1 and s_2, respectively, are connected in series, with the end A attached to a rigid support and the end C loaded by a force W. The calculation of spring deflections by the conventional approach is trivial. However, a finite element analysis of the problem will permit the essential features of the method to be demonstrated in a straightforward manner.

Application of the method can be broken down into a number of steps which are similar, whatever the problem.

5.2.1 Discretisation
We are dealing here with a problem which is discrete by nature. The choice of element type and discretisation are, therefore, predetermined; each spring is

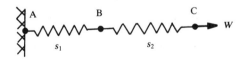

Fig. 5.2—Linear spring problem.

represented by a single element. The typical spring element e shown in Fig. 5.3 has a spring stiffness s_e and the two nodal points i and j. *Nodal point displacements* are denoted by δ_i, δ_j and *nodal point forces* by F_i^e, F_j^e. The forces carry a superscript e because nodes i and j are common to more than one element, each of which will make a distinct contribution. Such a distinction is not required for a nodal displacement which must have the same value for all elements connected to the particular node.

5.2.2 Element stiffness matrix

The next task is to derive the characteristics of the typical element by obtaining relationships between the nodal point forces and displacements. Putting the increase in length of element e as

$$\Delta_e = \delta_j - \delta_i \tag{5.2}$$

we can write equation (5.1) for the element spring force as

$$P_e = s_e(\delta_j - \delta_i) . \tag{5.3}$$

The forces at nodes i and j are in equilibrium if

$$\begin{aligned} F_i^e &= -P_e \\ F_j^e &= P_e \end{aligned} \tag{5.4}$$

and, after substituting for P_e from (5.3), we can express these in matrix notation as

$$\begin{bmatrix} F_i^e \\ F_j^e \end{bmatrix} = \begin{bmatrix} s_e & -s_e \\ -s_e & s_e \end{bmatrix} \begin{bmatrix} \delta_i \\ \delta_j \end{bmatrix} \tag{5.5}$$

or simply

$$[F^e] = [K^e][\delta^e]$$

where $[K^e]$ is termed the *element stiffness matrix*. The column vectors $[F^e]$ and $[\delta^e]$ contain the nodal point forces and displacements, respectively. Note that $[K^e]$ is a *symmetric matrix*.

5.2.3 System stiffness matrix

With the behaviour of a typical element established, we can now deduce the characteristics of the system of elements needed to represent the problem. The system consists of two elements, as shown in Fig. 5.4, where F_1, F_2 and F_3 denote the *externally applied forces* at nodes 1, 2 and 3. Free-body diagrams of each element are shown in Fig. 5.5.

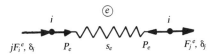

Fig. 5.3—Typical spring element.

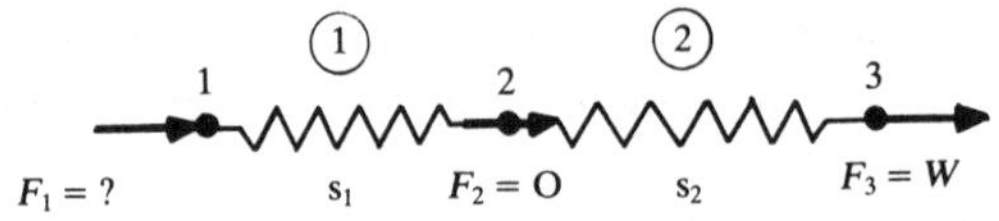

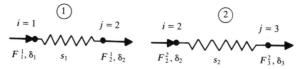

Fig. 5.4—System of linear spring elements.

Fig. 5.5—Free-body diagrams of linear spring elements.

At each node in the system there must be a balance between external and internal forces such that

$$F_1 = F_1^1$$
$$F_2 = F_2^1 + F_2^2 \tag{5.6}$$
$$F_3 = F_3^2$$

There are two contributions, F_2^1 and F_2^2, to the resultant force at node 2, because this node is common to elements 1 and 2. We can write out equations (5.5) for elements 1 and 2 as

$$\begin{bmatrix} F_1^1 \\ F_2^1 \end{bmatrix} = \begin{bmatrix} s_1 & -s_1 \\ -s_1 & s_1 \end{bmatrix} \begin{bmatrix} \delta_1 \\ \delta_2 \end{bmatrix} \quad , \quad \begin{bmatrix} F_2^2 \\ F_3^2 \end{bmatrix} = \begin{bmatrix} s_2 & -s_2 \\ -s_2 & s_2 \end{bmatrix} \begin{bmatrix} \delta_2 \\ \delta_3 \end{bmatrix} \tag{5.7}$$

which when substituted into (5.6) give

$$\begin{bmatrix} F_1 \\ F_2 \\ F_3 \end{bmatrix} = \begin{bmatrix} s_1 & -s_1 & 0 \\ -s_1 & s_1 + s_2 & -s_2 \\ 0 & -s_2 & s_2 \end{bmatrix} \begin{bmatrix} \delta_1 \\ \delta_2 \\ \delta_3 \end{bmatrix} \tag{5.8}$$

or simply

$$[F] = [K][\delta]$$

where $[F]$ and $[\delta]$ are the column vectors of nodal point forces and displacements, respectively, for the system of two elements. $[K]$ is referred to as the *system stiffness matrix* and is of order 3 for a system having three degrees of freedom, the unknown nodal displacements δ_1, δ_2 and δ_3.

A simple algorithm for the assembly of $[K]$ can be derived by first noting that $[F]$ is expressible as the summation of the contributions made by each element in the system. This is seen if we rewrite equations (5.6) as

$$
\begin{bmatrix} F_1 \\ F_2 \\ F_3 \end{bmatrix} = \begin{bmatrix} F_1^1 \\ F_2^1 \\ F_3^1 \end{bmatrix} + \begin{bmatrix} F_1^2 \\ F_2^2 \\ F_3^2 \end{bmatrix} = [F^1] + [F^2]
$$

where it is understood that $F_i^e = 0$ when node i does not belong to element e. If we substitute for $[F^1]$ and $[F^2]$ from (5.7), these become

$$
\begin{bmatrix} F_1 \\ F_2 \\ F_3 \end{bmatrix} = \begin{bmatrix} s_1 & -s_1 & 0 \\ -s_1 & s_1 & 0 \\ 0 & 0 & 0 \end{bmatrix} \begin{bmatrix} \delta_1 \\ \delta_2 \\ \delta_3 \end{bmatrix} + \begin{bmatrix} 0 & 0 & 0 \\ 0 & s_2 & -s_2 \\ 0 & -s_2 & s_2 \end{bmatrix} \begin{bmatrix} \delta_1 \\ \delta_2 \\ \delta_3 \end{bmatrix}
$$

$$
= [K^1][\delta] + [K^2][\delta]
$$

where the element stiffness matrices $[K^1]$ and $[K^2]$ have been expanded to the size of the system stiffness matrix (3 by 3) by the insertion of rows and columns of zeros in appropriate positions. It follows that

$$
[F] = [K][\delta]
$$

where

$$
[K] = \sum_{e=1}^{n_e} [K^e] \tag{5.9}
$$

and n_e denotes the number of elements in the system. Equation (5.9) provides a simple but yet powerful algorithm for assembling $[K]$, without the need to write out explicitly the equations of equilibrium for each node in turn.

5.2.4 Boundary conditions

The system stiffness matrix, as assembled using (5.9), is found to be *singular*, implying that the set of algebraic equations (5.8) do not have a unique solution for the nodal displacements. Since rigid-body displacement of the system has not been prevented, this is to be expected. To eliminate rigid-body displacement, we introduce the boundary condition $\delta_1 = 0$. Of the three externally applied nodal point forces, F_1 is an unknown support reaction, $F_2 = 0$ and F_3 has the prescribed value W. We can rewrite equations (5.8) to include these conditions as

$$
\begin{bmatrix} F_1 \\ 0 \\ W \end{bmatrix} = \begin{bmatrix} s_1 & -s_1 & 0 \\ -s_1 & s_1 + s_2 & -s_2 \\ 0 & -s_2 & s_2 \end{bmatrix} \begin{bmatrix} 0 \\ \delta_2 \\ \delta_3 \end{bmatrix} \tag{5.10}
$$

5.2.5 Solution
The set of equations (5.10) reduces to the two equations in δ_2 and δ_3 given by

$$\begin{bmatrix} 0 \\ W \end{bmatrix} = \begin{bmatrix} s_1 + s_2 & -s_2 \\ -s_2 & s_2 \end{bmatrix} \begin{bmatrix} \delta_2 \\ \delta_3 \end{bmatrix}$$

whose solution is $\delta_2 = W/s_1$ and $\delta_3 = W(1/s_1 + 1/s_2)$. The support reaction is given by the first of equations (5.10) as $F_1 = -s_1\delta_2 = -W$.

5.2.6 Element forces
With the displacements known, we can now calculate the spring forces in each element using (5.3) to give

$$P_1 = s_1(\delta_2 - \delta_1) = W$$
$$P_2 = s_2(\delta_3 - \delta_2) = W$$

The finite element solution of the problem is now complete and is *exact*. This follows from the use of an element whose characteristics model precisely those of the chosen subdivisions of the structure.

5.3 COMPUTATIONAL TECHNIQUES

For all but the simplest problems, finite element analysis is best implemented by a computer program which both assembles and solves the set of equilibrium equations. To introduce some of the computational techniques used, we now describe a FORTRAN 77 program FIESTA1 (the program name originates from FInite Element STress Analysis) for solving one-dimensional elastic problems. A complete listing of the source code is given in Appendix A2 together with a glossary of the FORTRAN variables used in Appendix A1. The program forms the basis of a more advanced version, FIESTA2, described in a later chapter, for the finite element analysis of two-dimensional problems.

The analysis of the deformation in an axially loaded elastic bar (see Fig. 5.6) typifies the type of problem that can be solved using FIESTA1. The bar has a

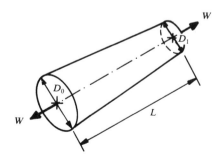

Fig. 5.6—Axially loaded elastic bar problem.

circular cross-section and a uniform taper which is sufficiently small for the deformation to be characterised by the axial displacement. Application of the finite element method involves essentially the same steps as those in the spring problem.

5.3.1 Discretisation

Unlike the previous example, this example is not discrete in nature but is a one-dimensional continuum problem, where it is possible to exercise choice in the selection of a suitable element type and in the subdivision into finite elements.

If we choose a slice of the bar defined by the two end nodes i and j as a typical element e (see Fig. 5.7), the task is then to derive the element stiffness matrix. The element has two degrees of freedom, the axial displacements δ_i and δ_j at the nodes, and there are two nodal point forces F_i^e and F_j^e. Using such elements, we subdivide the bar into a system of n_e elements and $n_n = n_e + 1$ nodes.

The subroutine SPEC is used to read the problem specification from a format-free data file on unit 5 (see Fig. 5.8). The variables D0 and D1 are the diameters of the end faces of the bar, FL is the length of the bar and E is the elastic

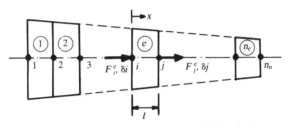

Fig. 5.7—Discretisation of an axially loaded bar.

```
D0, D1, FL, E
NONE, NODN, NOCO
NOE, NONS
NOPF
IPF(1,1), IPF(1,2), PF(1)
    .
    .
    .
IPF(NOPF,1), IPF(NOPF,2), PF(NOPF)
NOPD
IPD(1,1), IPD(1,2), PD(1)
    .
    .
    .
IPD(NOPD,1), IPD(NOPD,2), PD(NOPD)
```

Fig. 5.8—FIESTA1 data file

modulus of the material. NONE=2 is the number of nodes per element, NODN=1 the number of degrees of freedom per node and NOCO=1 the number of coordinates required to specify the element geometry. NODE=2 is the number of degrees of freedom per element. NOE is the number of elements, NONS is the number of nodes and NODS is the number of degrees of freedom in the system. XY(NS,1) is the axial coordinate of node NS. In later chapters, we shall use two-dimensional elements for which NOCO=2 and each node has two coordinates XY(NS,1) and XY(NS,2). IJ(IE,1) and IJ(IE,2) are the nodal indices i and j, respectively, for element IE.

NOPF is the number of prescribed forces, IPF(IBC,1) the node number and IPF(IBC,2)=1 the nodal degree of freedom to which the force boundary condition IBC relates. Elements used in later chapters have NODN=2 or 3. PF(IBC) is the prescribed magnitude of the force.

NOPD is the number of prescribed displacements, IPD(IBC,1) the node number and IPD(IBC,2)=1 the nodal degree of freedom to which the displacement boundary condition IBC relates. PD(IBC) is the prescribed magnitude of the displacement.

NOEQ is the number of equations to be solved. Later, we shall employ a method for introducing prescribed displacements which leads to a value for NOEQ which is less than NODS.

The problem specification is listed to a suitable device via unit 6 by subroutine LIST.

5.3.2   Element stiffness matrix

The axial displacement $u(x)$ within a typical element e of length l can be approximated by the *polynomial function*

$$u(x) = C_0 + C_1 x + C_2 x^2 + \ldots + C_n x^n$$

where the coefficients $C_0, C_1 \ldots$ must be evaluated in terms of the nodal degrees of freedom. Since each element has only two degrees of freedom, δ_i and δ_j, it follows that the polynomial can only contain two terms. If we put $C_i = 0$ for $i = 2, 3, \ldots,$ then $u(x)$ reduces to the *linear function*

$$u(x) = C_0 + C_1 x \tag{5.11}$$

The values of $u(x)$ at the nodes will match the nodal displacements if $u(0) = \delta_i$ and $u(l) = \delta_j$. These conditions are satisfied if

$$C_0 = \delta_i/l \quad \text{and} \quad C_1 = (\delta_j - \delta_i)/l$$

and thus (5.11) becomes

$$u(x) = [N_i \quad N_j] \begin{bmatrix} \delta_i \\ \delta_j \end{bmatrix}$$

or

$$[u] = [N][\delta^e] \tag{5.12}$$

where

$$N_i(x) = 1 - x/l \quad \text{and} \quad N_j(x) = x/l$$

are the *shape functions* which are used to interpolate the value of $[u] = u(x)$ in terms of the nodal displacements.

The axial strain $e(x) = du/dx$ found using (5.12) is given by

$$e(x) = \left[\frac{dN_i}{dx} \quad \frac{dN_j}{dx} \right] \begin{bmatrix} \delta_i \\ \delta_j \end{bmatrix}$$

or

$$[e] = [B][\delta^e] \tag{5.13}$$

where $[B] = [-1/l \quad 1/l]$. It follows that the strain is *constant* within each element.

If the material obeys Hooke's law, then the axial stress $[\sigma] = \sigma(x)$ is given by

$$[\sigma] = [D][e] \tag{5.14}$$

where $[D] = E$, the elastic modulus of the material. Substituting for $[e]$ from (5.13) into (5.14), we obtain

$$[\sigma] = [D][B][\delta^e] \tag{5.15}$$

In two-dimensional problems examined in later chapters, $[u]$, $[\sigma]$ and $[e]$ are column vectors and $[D]$ is a symmetric matrix. In the present problem, they are all scalars.

The nodal forces can be expressed in terms of the element stress as

$$\begin{bmatrix} F_i^e \\ F_j^e \end{bmatrix} = A \begin{bmatrix} -1 \\ 1 \end{bmatrix} [\sigma]$$

where A is the cross-section area of the bar half-way along its length. If we substitute for $[\sigma]$ from (5.15), these become

$$[F^e] = [K^e][\delta^e] \tag{5.16}$$

where

$$[K^e] = V[B]^{\mathrm{T}}[D][B]$$

is the element stiffness matrix and V is the volume of the element. Substituting for $[B]$ and $[D]$, we have finally

$$[K^e] = \frac{EA}{l} \begin{bmatrix} 1 & -1 \\ -1 & 1 \end{bmatrix} \tag{5.17}$$

Comparing this with the stiffness matrix for the spring element given in (5.5), we
see that the present element is equivalent to a spring of stiffness $s_e = EA/l$.

Subroutine ELEMNT evaluates the element stiffness array SE according to
(5.17). The variables I and J are the nodal indices i and j for element IE and FLE is
the element length l. XMEAN is the x coordinate of the midpoint of the element
and DMEAN is the diameter and A the cross-section area A at this point. S is the
equivalent spring stiffness s_e.

5.3.3 System stiffness matrix

The algorithm given by (5.9) forms the basis of the procedure for assembling $[K]$.
Before it can be implemented, however, it is necessary to expand each element
stiffness matrix to *system size*. In effect, this requires the relationship between row
and column positions of coefficients K_{rc}^e in $[K^e]$ and K_{rc} in $[K]$ to be established.

The set of equilibrium equations for the system of $n = n_n$ nodal points is of the
form

$$
\begin{bmatrix} F_1 \\ F_2 \\ \cdot \\ \cdot \\ F_i \\ \cdot \\ F_j \\ \cdot \\ \cdot \\ F_n \end{bmatrix}
=
\begin{bmatrix}
K_{11} & K_{12} & \cdot & \cdot & \cdot & \cdot & \cdot & \cdot & \cdot & \cdot \\
K_{21} & \cdot & \cdot & \cdot & \cdot & \cdot & \cdot & \cdot & \cdot & \cdot \\
\cdot & \cdot & \cdot & \cdot & \cdot & \cdot & \cdot & \cdot & \cdot & \cdot \\
\cdot & \cdot & \cdot & \cdot & \cdot & \cdot & \cdot & \cdot & \cdot & \cdot \\
\cdot & \cdot & \cdot & \cdot & K_{ii} & \cdot & K_{ij} & \cdot & \cdot & \cdot \\
\cdot & \cdot & \cdot & \cdot & \cdot & \cdot & \cdot & \cdot & \cdot & \cdot \\
\cdot & \cdot & \cdot & \cdot & K_{ji} & \cdot & K_{jj} & \cdot & \cdot & \cdot \\
\cdot & \cdot & \cdot & \cdot & \cdot & \cdot & \cdot & \cdot & \cdot & \cdot \\
\cdot & \cdot & \cdot & \cdot & \cdot & \cdot & \cdot & \cdot & \cdot & \cdot \\
\cdot & \cdot & \cdot & \cdot & \cdot & \cdot & \cdot & \cdot & \cdot & K_{nn}
\end{bmatrix}
\begin{bmatrix} \delta_1 \\ \delta_2 \\ \cdot \\ \cdot \\ \delta_i \\ \cdot \\ \delta_j \\ \cdot \\ \cdot \\ \delta_n \end{bmatrix}
\qquad (5.18)
$$

column c … i … j … ; row r … i … j

where i and j are the nodal point indices of a typical element e, with degrees of
freedom δ_i and δ_j. The resultant nodal forces F_i and F_j are defined by the ith and jth
equations, respectively, and degrees of freedom 1 and 2 for the element become
degrees of freedom i and j for the system. It follows that the element coefficients
K_{11}^e and K_{12}^e which determine the contribution F_i^e are to be summed with the system
coefficients K_{ii} and K_{ij}, respectively. Similarly, K_{21}^e and K_{22}^e, which determine the
contribution F_j^e, are to be summed with K_{ji} and K_{jj}, respectively. The assembly
algorithm, applied to element e, can now be written as

$$
\begin{aligned}
K_{ii}' &= K_{ii} + K_{11}^e, & K_{ij}' &= K_{ij} + K_{12}^e \\
K_{ji}' &= K_{ji} + K_{21}^e, & K_{jj}' &= K_{jj} + K_{22}^e
\end{aligned}
\qquad (5.19)
$$

where the remaining system stiffness coefficients are unaltered in value. This summation is repeated for each element in turn.

The fully assembled stiffness matrix for a system of four elements is

$$[K] = \begin{bmatrix} s_1 & -s_1 & 0 & 0 & 0 \\ -s_1 & s_1 + s_2 & -s_2 & 0 & 0 \\ 0 & -s_2 & s_2 + s_3 & -s_3 & 0 \\ 0 & 0 & -s_3 & s_3 + s_4 & -s_4 \\ 0 & 0 & 0 & -s_4 & s_4 \end{bmatrix}$$

$$\overleftarrow{} b = 2 \overrightarrow{}$$

where the broken boxes indicate the individual element stiffness matrices, and their overlap shows the coupling between them. It is evident that $[K]$ is a *symmetric tridiagonal* matrix with a *semi-bandwidth* $b = 2$. For a large system of elements a high proportion of the stiffness coefficients is equal to zero. To exploit this fact, we store $[K]$ as an n by 2 array SS, as shown in Fig. 5.9. SS stores only those non-zero coefficients which lie on, or above, the leading diagonal of $[K]$. When a coefficient K_{rc} lying below the leading diagonal is required, the symmetric coefficient K_{cr} is used in its place. The equivalent row, column positions in the matrix $[K]$ and the array SS are summarised in Table 5.1.

The assembly algorithm (5.19) is implemented in subroutine ASSEM where elements of the array SS and the nodal forces array FS are first zeroed. For each element IE the element stiffness array SE is determined in subroutine ELEMNT. Array IRC defines, for each row, column position IRE,ICE in $[K^e]$, the

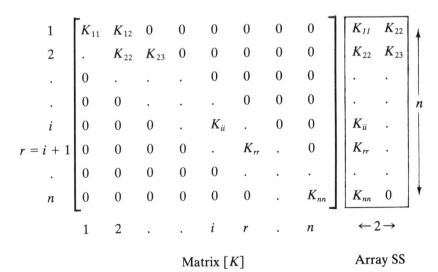

Fig. 5.9—Storage scheme for a symmetric tridiagonal matrix.

Table 5.1 Equivalent row, column positions in a symmetric banded matrix $[K]$ and the FORTRAN array SS

Matrix $[K]$	FORTRAN array SS	
	$c \geqslant r$	$c < r$
Row r	r	c
Column c	$c - r + 1$	$r - c + 1$

corresponding position in $[K]$. The equivalent position IRS, ICS in the array SS is obtained using Table 5.1. The summation for $[K]$ is then performed according to (5.19) only for those stiffness coefficients stored in SS. Finally, any prescribed nodal point forces are summed with the corresponding elements of the array FS, using IDS as a pointer. IDS is the system degree of freedom number to which each force boundary condition IBC relates.

5.3.4 Boundary conditions
A convenient method for imposing the displacement boundary condition $\delta_i = \alpha_i$ is to multiply the diagonal coefficient K_{ii} by a very large number, say 10^{20}, and to replace the nodal force F_i by $\alpha_i K_{ii} 10^{20}$. Equation i is thereby modified to

$$K_{i,i-1}\delta_{i-1} + K_{ii}10^{20}\delta_i + K_{i,i+1}\delta_{i+1} = \alpha_i K_{ii}10^{20} \tag{5.20}$$

The second term on the left-hand side of the equation now dominates that side and (5.20) is, therefore, very nearly equivalent to the required condition. Neither the bandwidth nor the symmetry of $[K]$ is affected by prescribing displacements in this way.

The boundary condition $\delta_1 = 0$ is applied using this method by putting $\alpha_1 = 0$.

5.3.5 Solution
The set of tridiagonal equations

$$\begin{bmatrix} K_{11} & K_{12} & 0 & 0 & 0 & 0 & 0 \\ K_{21} & K_{22} & K_{23} & 0 & 0 & 0 & 0 \\ 0 & K_{32} & K_{33} & K_{34} & 0 & 0 & 0 \\ 0 & 0 & . & . & . & 0 & 0 \\ 0 & 0 & 0 & . & . & . & 0 \\ 0 & 0 & 0 & 0 & . & . & . \\ 0 & 0 & 0 & 0 & 0 & . & K_{nn} \end{bmatrix} \begin{bmatrix} \delta_1 \\ \delta_2 \\ \delta_3 \\ . \\ . \\ . \\ \delta_n \end{bmatrix} = \begin{bmatrix} F_1 \\ F_2 \\ F_3 \\ . \\ . \\ . \\ F_n \end{bmatrix} \tag{5.21}$$

can be solved by various direct or iterative methods. The method adopted here is the method of *Gaussian elimination* in which $[K]$ is first reduced to an upper triangular matrix after which the solution is obtained by back substitution.

We start with the elimination of δ_1 from the second of equations (5.21) by subtracting from that equation a suitable multiple R of the first equation. The coefficients of the second equation are thereby redefined as

$$K'_{21} = K_{21} - K_{11}R = 0 \qquad \text{if } R = K_{21}/K_{11}$$

$$K'_{22} = K_{22} - K_{12}R$$

$$K'_{23} = K_{23} - 0 \times R = K_{23}$$

$$F'_2 = F_2 - F_1R$$

The new second equation is then used to eliminate δ_2 from the third equation whose coefficients are thereby redefined as

$$K'_{32} = K_{32} - K_{22}R = 0 \qquad \text{if } R = K_{32}/K_{22}$$

$$K'_{33} = K_{33} - K_{23}R$$

$$K'_{34} = K_{34} - 0 \times R = K_{34}$$

$$F'_3 = F_3 - F_2R$$

After the elimination of δ_i from equation $i + 1$, using equation i, the coefficients of the former become

$$K'_{i+1,i} = K_{i+1,i} - K_{ii}R = 0 \qquad \text{if } R = K_{i+1,i}/K_{ii}$$

$$K'_{i+1,i+1} = K_{i+1,i+1} - K_{i,i+1}R$$

$$K'_{i+1,i+2} = K_{i+1,i+2} - 0 \times R = K_{i+1,i+2} \tag{5.22}$$

$$F'_{i+1} = F_{i+1} - F_iR$$

The process continues until $i = n - 1$.

When forward reduction is complete, equations (5.21) take the form

$$
\begin{bmatrix}
K_{11} & K_{12} & 0 & 0 & 0 & 0 & 0 \\
0 & K'_{22} & K'_{23} & 0 & 0 & 0 & 0 \\
0 & 0 & K'_{33} & K'_{34} & 0 & 0 & 0 \\
0 & 0 & 0 & . & . & 0 & 0 \\
0 & 0 & 0 & 0 & . & . & 0 \\
0 & 0 & 0 & 0 & 0 & . & . \\
0 & 0 & 0 & 0 & 0 & 0 & K'_{nn}
\end{bmatrix}
\begin{bmatrix}
\delta_1 \\
\delta_2 \\
\delta_3 \\
. \\
. \\
. \\
\delta_n
\end{bmatrix}
=
\begin{bmatrix}
F_1 \\
F'_2 \\
F'_3 \\
. \\
. \\
. \\
F'_n
\end{bmatrix}
\tag{5.23}
$$

From equation n, it follows that

$$\delta_n = \frac{F'_n}{K'_{nn}} \tag{5.24}$$

and from equation $n - 1$ that

$$\delta_{n-1} = \frac{(F'_n - K'_{n-1,n}\delta_n)}{K'_{n-1,n-1}}$$

In this way, we reach the solution for δ_i from equation i given by

$$\delta_i = \frac{(F'_i - K_{i,i+1}\delta_{i+1})}{K'_{ii}} \tag{5.25}$$

and so on, until $i = 1$.

Gaussian elimination for a set of symmetric tridiagonal equations is performed by subroutine SOLVE. Equations (5.22), (5.24) and (5.25), which define the process in terms of the coefficients of the matrix $[K]$, are converted into algorithms involving the elements of the array SS using the row, column relationships defined in Table 5.1. The stiffness coefficient $K_{i+1,i}$ used to calculate R for the reduction step i, is replaced by $K_{i,i+1}$ which corresponds to the array element SS(I,2). Division by zero in the calculation of R, is prevented by testing the ratio K_{ii}/K^*_{ii} where K^*_{ii} is the original value of the diagonal coefficient stored in the array DS. A zero value for this ratio implies that $[K]$ is a singular matrix and that rigid-body displacements have not been prevented by properly specifying the displacement boundary conditions. The test has another purpose as we shall see later.

During back substitution the array FS is re-used to store the solution for the nodal displacements, thereby avoiding the unnecessary use of additional memory.

5.3.6   Element stresses

The axial stress σ in element e is given by (5.15) as

$$[\sigma] = [D][B][\delta^e]$$
$$= \frac{E}{l}[-1 \quad 1]\begin{bmatrix} \delta_i \\ \delta_j \end{bmatrix} \tag{5.26}$$
$$= \frac{E}{l}(\delta_j - \delta_i)$$

and is calculated in subroutine STRESS where FS(I) and FS(J) are now the nodal displacements for element IE. The array SS is re-used to store the values of the stresses, and SS(IE,1) is now the axial stress in element IE.

Subroutine OUTPUT lists the solution for the nodal displacements and element stresses.

5.4 NUMERICAL EXAMPLE

To verify the program, we first solve a problem where the finite element solution is exact. This is the special case $D_0 = D_1$ (see Fig. 5.6), where the strain is *uniform* along the length of the bar, and therefore in each element, as assumed in the formulation of the element stiffness matrix.

We now use the program to solve the problem for $D_0 = L/5$ and $D_1 = L/10$ where the exact solution for the total elongation of the bar is given by $\Delta L = 4WL/\pi D_0 D_1 E$. This solution can be expressed in dimensionless form (see section 4.7) as

$$\frac{\Delta L\ EL}{W} = \frac{4L^2}{\pi D_0 D_1} = \frac{200}{\pi}$$

In this case, there is a *discretisation error* in the finite element solution for ΔL resulting from the use of a discrete model which is not an exact representation of the bar.

The discretisation error is given in Table 5.2 as a function of the element length l and is almost proportional to l^2. The *convergence rate* is, therefore, said to be *of the order l^2*. If $\Delta L^{(1)}$ and $\Delta L^{(2)}$ are solutions obtained using elements of length l and $l/2$, respectively, then

$$\frac{\Delta L - \Delta L^{(1)}}{\Delta L - \Delta L^{(2)}} \approx \frac{l^2}{(l/2)^2} = 4$$

Hence we can obtain an improved approximation from the *extrapolation formula*

$$\Delta L = \frac{4\ \Delta L^{(2)} - \Delta L^{(1)}}{3} \tag{5.27}$$

As an illustration we take $\Delta L^{(1)}$ and $\Delta L^{(2)}$ corresponding to $l = L/5$ and $l = L/10$, respectively, for which (5.27) gives $\Delta L\ EL/W = 63.660$. This is actually a better approximation than that obtained using $l = L/50$!

It is significant that the finite element solution *underestimates* ΔL and therefore gives a higher axial stiffness for the bar than the exact solution does. This is a

Table 5.2 Finite element solution for the
elongation ΔL of an axially loaded tapered bar

l	n_e	$\Delta L\ EL/W$	Error (%)
L	1	56.588	−11.1
$L/2$	2	61.532	− 3.35
$L/5$	5	63.296	− 0.575
$L/10$	10	63.570	− 0.145
$L/20$	20	63.638	− 0.0364
$L/50$	50	63.658	− 0.00584
Exact solution		63.662	

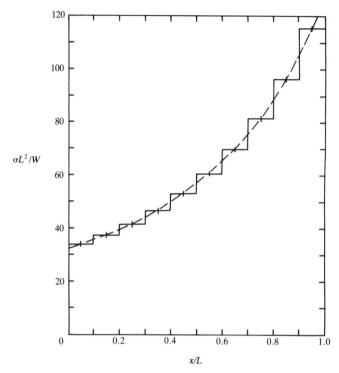

$\sigma L^2/W$

Fig. 5.10—Finite element solution (——) for the axial stress variation when $l = L/10$ and
$n_e = 10$. The exact solution (– – –) is also shown.

characteristic feature of the finite element formulation which we shall discuss more
fully in the next chapter.

The solution for the stresses when $l = L/10$ is given in dimensionless form in
Fig. 5.10. The stresses are *uniform* within each element, and are *exact* at the
midpoint of each element. The maximum errors occur at the nodal points where
there are *step discontinuities* between adjacent elements.

5.5   COMPUTATIONAL ERRORS

We are already familiar with the fact that, in general, a finite element solution is
only approximate. The discretisation error arises from the use of a discrete model
which is seldom an exact representation of the physical problem. This error may be
reduced to an acceptable level by employing a sufficiently refined mesh of
elements.

There is another quite distinct source of error in finite element analysis which is
associated with the way in which computers store and manipulate numbers. Most
computers store numbers internally in a *binary* form using a fixed number of binary
digits or *bits*. For an *integer number*, this means that there is a limit on the *range* of
numbers that can be represented. In the case of a *floating point number* there is, in
addition, a limit on the *accuracy* with which the number can be represented.

The binary representation of floating point numbers uses a fixed number of bits for both the fractional part, or *mantissa*, and the *exponent*. Many FORTRAN 77 compilers have a default allocation of 4 bytes for a *single precision* real variable: one for the exponent and three for the mantissa, including one *sign bit* in each case (see Fig. 5.11). The exponent is therefore limited to the range ±127 whilst the mantissa must be *rounded* to the nearest value that can be stored exactly in the 23 bits available. This latter constraint means that the accuracy with which a floating point number can be represented is limited to about seven decimal digits.

Round-off error can grow unacceptably large during a sequence of floating point arithmetic operations, such as that employed in Gaussian elimination. The problem is most apparent in the subtraction of two numbers of nearly equal magnitude. Suppose the numbers have exact values x_1 and x_2 and are represented in binary as $x_1(1 + e_1)$ and $x_2(1 + e_2)$, where e_1 and e_2 are the round-off errors for the decimal-to-binary conversions. For the difference $x_1 - x_2$ the round-off error is given by

$$e_d = \frac{x_1}{x_1 - x_2} e_1 - \frac{x_2}{x_1 - x_2} e_2 \tag{5.28}$$

in which the coefficients of e_1 and e_2 are both greater than unity. It follows that the errors e_1 and e_2 are magnified in the computed difference—the smaller the difference $x_1 - x_2$, the greater is this magnification.

The dangers of excessive round-off error in Gaussian elimination can be demonstrated by a numerical example of the solution to the linear spring problem shown in Fig. 5.2. The equilibrium equations (5.10) reduce to

$$\begin{bmatrix} s_1 + s_2 & -s_2 \\ -s_2 & s_2 \end{bmatrix} \begin{bmatrix} \delta_2 \\ \delta_3 \end{bmatrix} = \begin{bmatrix} 0 \\ W \end{bmatrix} \tag{5.29}$$

and, to solve these by elimination, we multiply the first equation by $s_2/(s_1 + s_2)$ and add it to the second to give

$$\begin{bmatrix} s_1 + s_2 & -s_2 \\ 0 & s_2 - s_2^2/(s_1 + s_2) \end{bmatrix} \begin{bmatrix} \delta_2 \\ \delta_3 \end{bmatrix} = \begin{bmatrix} 0 \\ W \end{bmatrix} \tag{5.30}$$

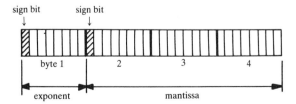

Fig. 5.11—Typical memory allocation for a single precision real variable in FORTRAN 77, using 4 bytes.

The second of equations (5.30) gives the solution for δ_3 as

$$\delta_3 = \frac{W(s_1 + s_2)}{s_2(s_1 + s_2) - s_2^2} \tag{5.31}$$

If the spring stiffness s_2 is *very large* in relation to s_1, the denominator of the right-hand side of (5.31) is a difference of two terms which will be nearly equal to one another. The round-off error in the computed value of the difference will be magnified in the manner indicated by (5.28) and the solution for both δ_2 and δ_3 will be in error.

This example typifies the physical situation conducive to large round-off errors where a stiff element is attached to elements with very much lower stiffnesses. In these cases the stiffness matrix $[K]$ is said to be *ill-conditioned* and the solution of the equilibrium equations is sensitive to small changes in the stiffness coefficients which occur when trailing digits are lost in rounding.

Ill-conditioning reveals itself during the forward reduction process by the appearance of a coefficient K_{ii} on the leading diagonal of $[K]$ after the reduction step $i - 1$ which is very small compared with its original value K_{ii}^*. This *diagonal decay* can be monitored by testing the ratio $\lambda = K_{ii}/K_{ii}^*$ before the start of each reduction step. If $\lambda = 10^{-n}$, then about n trailing digits of K_{ii}^* have been lost during the subtractions involved in the calculation of K_{ii}. The solution is abandoned if λ is less than, say, 10^{-3}. This test will also detect, as we saw earlier, the extreme case of ill-conditioning which occurs when $[K]$ is singular, and $\lambda = 0$.

To improve the accuracy with which floating-point numbers can be represented, *double precision* real variables can be employed. Here the mantissa is allocated double the number of bytes used in single precision and results in an accuracy of about 14 decimal digits. Not only are the memory requirements of a program thereby increased but so too is the execution time, because of the additional complexity of double precision arithmetic.

5.6 CONCLUSION

In this chapter the application of the finite element method to one-dimensional problems has been described. On the basis of an assumed displacement field for each bar element, we derived the set of nodal force–displacement equations. This procedure typifies the *displacement method* of finite element analysis in which it is the nodal displacements which are the primary unknowns. The element stresses are deduced from the solution for the displacements.

An alternative procedure, the *equilibrium method*, is possible in which the stresses are the primary unknowns. In the *mixed* or *hybrid method*, both stresses and displacements are primary unknowns.

We have been able to derive element stiffness matrices by a direct application of the equilibrium conditions. For one-dimensional problems, such a *direct formulation* is often adequate and has a readily appreciated physical interpretation. A more versatile formulation, based on the principle of virtual work, will be outlined in the next chapter.

PROBLEMS

5.1   A set of linear springs is connected as shown. The left-hand end of spring s_1 is connected to a rigid support while the right-hand end of spring s_4 carries a load W. Assemble the system stiffness matrix and solve for the nodal displacements and spring forces when $s_1 = s_2 = s_3 = s_4 = s$.

(Answers: 0, W/s, $3W/2s$, $5W/2s$; W, $W/2$, $W/2$, W.)

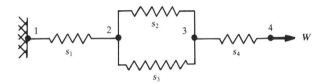

5.2   An elastic rod ABCD has a cross-section of diameter 10 mm over the portions AB and CD and 15 mm over the portion BC. The rod ends A and D are rigidly constrained, and equal and opposite axial loads $W = 10$ kN are applied at B and C. Use the finite element method to calculate the axial stresses in the rod assuming $E = 200$ GN/m^2.

(Answers: 31.8 MN/m^2, -42.4 MN/m^2.)

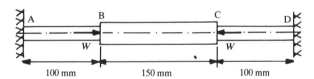

5.3   A tapered elastic bar element of circular cross-section is subjected to torsional loading about its axis. T_i^e, T_j^e denote the torsional loads at the nodes, and ϕ_i, ϕ_j the corresponding angles of rotation. Show that the equilibrium equations for the element are

$$\begin{bmatrix} T_i^e \\ T_j^e \end{bmatrix} = \frac{GJ}{l} \begin{bmatrix} 1 & -1 \\ -1 & 1 \end{bmatrix} \begin{bmatrix} \phi_i \\ \phi_j \end{bmatrix}$$

where G is the shear modulus of the material, J is the polar second moment of area of the middle cross-section of the element and l is the length of the element.

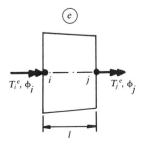

5.4   A stepped elastic shaft AB carries a torsional load T at the end B. Use the finite element method to calculate the angle of rotation at B if the end A is rigidly constrained. The material has an elastic shear modulus G and the polar second moments of area of the three sections are J_1, J_2 and J_3.

(Answers: $T(L_1/J_1 + L_2/J_2 + L_3/J_3)/G$.)

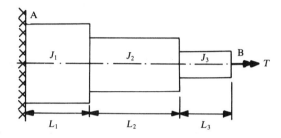

Chapter 6

Virtual Work Approach To Finite Element Analysis

6.1 INTRODUCTION

In the previous chapter, we saw how the set of equilibrium equations $[F] = [K][\delta]$ can be assembled for a system of one-dimensional bar elements using the direct formulation of the finite element method. An alternative approach is required, however, for more complicated elements, and where the external loading includes surface and body forces, in addition to point forces at the nodes. The *principle of virtual work* provides the basis for such an approach.

6.2 PRINCIPLE OF VIRTUAL WORK FOR A DISCRETE PROBLEM

We begin by deriving the virtual work principle for the linear spring problem of section 5.2, where a typical spring element is shown in Fig. 5.3. If the nodal forces F_i^e and F_j^e, together with the spring force P_e fulfil the requirements of *equilibrium*, it was shown that

$$
\begin{aligned}
F_i^e &= -P_e \\
F_j^e &= P_e
\end{aligned}
\tag{5.4}
$$

For the nodal displacements δ_i and δ_j, together with the spring deflection Δ_e, to fulfil the requirements of *compatibility*, it was shown that

$$
\Delta_e = \delta_j - \delta_i
\tag{5.2}
$$

If we multiply the first of equations (5.4) by δ_i, the second by δ_j, add and use (5.2), we obtain

$$
F_i^e \delta_i + F_j^e \delta_j = P_e \Delta_e
\tag{6.1}
$$

This equation is a statement of the principle of virtual work for a single element.

For a system of n_e elements and n_n nodes, both sides of (6.1) are summed over the system to give

$$\sum_{n=1}^{nn} F_n \delta_n = \sum_{e=1}^{ne} P_e \Delta_e \tag{6.2}$$

To have a proper understanding of the principle, it is important to emphasise that for equations (6.1) and (6.2) to be valid it is not necessary for the set of forces (F_n, P_e) to be the actual forces in the system. Neither is it necessary for the displacements (δ_n, Δ_e) to be the actual displacements. The two conditions that must be satisfied, however, are the following.

(1) The set of forces (F_n, P_e) satisfy all the requirements of equilibrium, including any boundary conditions.
(2) The set of displacements (δ_n, Δ_e) satisfy all the requirements of compatibility, including any boundary conditions.

Provided that these conditions are fulfilled, then (F_n, P_e) and (δ_n, Δ_e) may be any hypothetical or *virtual* sets.

In the derivation of the principle, no reference was made to the load–deflection relationship for the springs, which may, therefore, be elastic or inelastic. It follows that the displacements need not be those produced by the forces, whether or not the latter are the actual forces.

Both sides of equations (6.1) and (6.2) consist of terms which are each products of forces and displacements. It is tempting, therefore, to interpret the principle as expressing the equality between the work done on the system by external forces, and the work absorbed in the springs by internal forces. In general, however, it is wrong to interpret the principle in this way. The proper definitions for the work done W_n by each external force F_n, and the work absorbed U_e by each internal force P_e are given by

$$W_n = \int F_n \, d\delta_n, \qquad U_e = \int P_e \, d\Delta_e \tag{6.3}$$

where the forces are assumed to be quasi-statically increased from initial zero values. The usual interpretation of work done as the area under the corresponding force–displacement curve is shown in Fig. 6.1 for each work term. Clearly the integrals in (6.3) are never equal to $F_n \delta_n$ and $P_e \Delta_e$ for any realistic curves.

Only in the special case of a *linear elastic* system, where

$$W_n = \tfrac{1}{2} F_n \delta_n, \qquad U_e = \tfrac{1}{2} P_e \Delta_e$$

does the principle reduce to a real work balance when factors of $\tfrac{1}{2}$ are introduced on either side of (6.2) to give

$$\sum_{n=1}^{nn} W_n = \sum_{e=1}^{ne} U_e$$

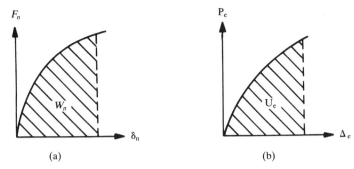

Fig. 6.1—(a) Work done by an external force F_n and (b) work absorbed by a spring force P_e.

Here the principle amounts to a statement of the *first law of thermodynamics* for the system. The applicability of the principle is very broad however, and the real work interpretation is best avoided.

To appreciate the use of the principle in finite element analysis, we consider its application to the problem of two spring elements shown in Fig. 5.4. The virtual work equation (6.1) for a single element e can be written as

$$[\delta^e]^T[F^e] = s_e[\delta^e]^T \begin{bmatrix} -1 \\ 1 \end{bmatrix} [-1 \quad 1][\delta^e]$$

which simplifies to

$$[F^e] = [K^e][\delta^e]$$

where

$$[K^e] = \begin{bmatrix} s_e & -s_e \\ -s_e & s_e \end{bmatrix}$$

is the element stiffness matrix, as derived previously using the direct formulation.

The principle can also be applied to the system of elements using (6.2). Since

$$\sum_{n=1}^{n_n} F_n \delta_n = [\delta]^T[F] \qquad \text{and} \qquad \sum_{e=1}^{n_e} P_e \Delta_e = [P]^T[\Delta]$$

where

$$[\Delta] = \begin{bmatrix} \Delta_1 \\ \Delta_2 \end{bmatrix} = \begin{bmatrix} -1 & 1 & 0 \\ 0 & -1 & 1 \end{bmatrix} \begin{bmatrix} \delta_1 \\ \delta_2 \\ \delta_3 \end{bmatrix} = [T][\delta]$$

and

$$[P] = \begin{bmatrix} P_1 \\ P_2 \end{bmatrix} = \begin{bmatrix} s_1 & 0 \\ 0 & s_2 \end{bmatrix} \begin{bmatrix} \Delta_1 \\ \Delta_2 \end{bmatrix} = [S][\Delta]$$

it follows that (6.2) can be expressed as

$$[\delta]^T[F] = [P]^T[\Delta] = [\Delta]^T[S]^T[\Delta]$$
$$= [\delta]^T[T]^T[S]^T[T][\delta]$$

or simply as

$$[F] = [K][\delta]$$

where

$$[K] = [T]^T[S]^T[T] = \begin{bmatrix} s_1 & -s_1 & 0 \\ -s_1 & s_1 + s_2 & -s_2 \\ 0 & -s_2 & s_2 \end{bmatrix}$$

is the system stiffness matrix, as given by equation (5.8).

From the above examples it is seen that the principle of virtual work provides an alternative means of deriving both element and system stiffness matrices. In finite element analysis, it is convenient to employ the principle in its above form only for the task of deriving $[K^e]$. The derivation of $[K]$ is best accomplished using the principle in the form expressed by the assembly algorithm given in equation (5.9).

6.3 THEOREM OF MINIMUM POTENTIAL ENERGY

We apply the principle once again to the system of elastic springs using the set of actual forces (F_n, P_e), as above. In place of the actual displacements, however, we now use a set of *virtual displacements* $(d\delta_n, d\Delta_e)$ which represent a *small* disturbance of the actual equilibrium displacements (δ_n, Δ_e). In this case the virtual work equation (6.2) becomes

$$\sum_{n=1}^{nn} F_n \, d\delta_n = \sum_{e=1}^{ne} P_e \, d\Delta_e \tag{6.4}$$

and, since the system is elastic, it is possible to express the internal forces as

$$P_e = \frac{\partial}{\partial \Delta_e} \left(\int P_e \, d\Delta_e \right) \tag{6.5}$$

After substituting for P_e from (6.5), we can write (6.4) as

$$d\left(\sum_{e=1}^{ne} \int P_e \, d\Delta_e - \sum_{n=1}^{nn} F_n \delta_n \right) = 0$$

or more concisely

$$d(U - V) = 0 \tag{6.6}$$

where

$$U = \sum_{e=1}^{ne} \int P_e \, d\Delta_e = \sum_{e=1}^{ne} U_e$$

is defined as the strain energy of the system, and

$$V = \sum_{n=1}^{nn} F_n \delta_n = \sum_{n=1}^{nn} V_n$$

is defined as the *potential energy of the external forces*. If the potential energy of the system is defined as

$$\Pi = U - V \qquad (6.7)$$

then the principle reduces to

$$d\Pi = 0 \qquad (6.8)$$

which represents a *stationary value* for the potential energy. It can further be shown that, for a system in *stable equilibrium*, Π is a *minimum*. Although we have derived (6.8) for a particular problem, it can in fact be shown to hold for any elastic structure. Thus, we have the *theorem of minimum potential energy* for an elastic structure, which may be stated as follows.

Of all the sets of compatible displacements which satisfy the boundary conditions, those which correspond to the actual forces make the potential energy of the system a minimum.

The theorem is an example of a *variational principle* of continuum mechanics involving the minimisation of a *functional*, the potential energy in this case.

6.4 PRINCIPLE OF VIRTUAL WORK FOR A SOLID CONTINUUM

A virtual work balance, of the form given by equation (6.2) for a discrete problem, is also valid for a solid continuum. The set of forces and displacements postulated in the principle now require the following generalised interpretations.

(1) Equilibrium forces: these are now the surface tractions $[t]$ per unit area applied to the surface S of the solid, the body forces $[b]$ per unit volume acting over the volume V and the internal stresses $[\sigma]$ given by

$$[t] = [t_x \quad t_y \quad t_z]^T, \qquad [b] = [b_x \quad b_y \quad b_z]^T,$$

and

$$[\sigma] = [\sigma_{xx} \quad \sigma_{yy} \quad \sigma_{zz} \quad \tau_{xy} \quad \tau_{yz} \quad \tau_{zx}]^T$$

Together these generalised forces must satisfy the requirements of equilibrium, both in the interior of the solid and on its surface.

(2) Compatible displacements: these are now the displacement components $[u]$ and the strain components $[e]$ at any point given by

$$[u] = [u \quad v \quad w]^{T}, \qquad [e] = [e_{xx} \quad e_{yy} \quad e_{zz} \quad \gamma_{xy} \quad \gamma_{yz} \quad \gamma_{zx}]^{T}$$

which must satisfy the requirements of compatibility, both in the interior and on the surface of the solid.

Using only these assumptions, and without reference to the material behaviour, the virtual work balance is

$$\int (ut_x + vt_y + wt_z) \, dS + \int (ub_x + vb_y + wb_z) \, dV$$
$$= \int (\sigma_{xx} e_{xx} + \sigma_{yy} e_{yy} + \sigma_{zz} e_{zz} + \tau_{xy} \gamma_{xy} + \tau_{yz} \gamma_{yz} + \tau_{zx} \gamma_{zx}) \, dV$$

or in matrix notation

$$\int [u]^{T}[t] \, dS + \int [u]^{T}[b] \, dV = \int [e]^{T}[\sigma] \, dV \qquad (6.9)$$

As before, it is not necessary for the set of either the forces or the displacements to be the actual ones present in the structure.

If we take the set of actual forces, and a set of virtual displacements du, dv, dw which represent a small disturbance of the actual equilibrium displacements, then for an elastic solid we can show that the theorem of minimum potential energy holds. For an elastic solid the strain energy (see section 3.3) is given by

$$U = \int (\sigma_{xx} \, de_{xx} + \sigma_{yy} \, de_{yy} + \sigma_{zz} \, de_{zz} + \tau_{xy} \, d\gamma_{xy} + \tau_{yz} \, d\gamma_{yz} + \tau_{zx} \, d\gamma_{zx}) \, dV$$

and the potential energy of the external forces is

$$V = \int (t_x \, du + t_y \, dv + t_z \, dw) \, dS + \int (b_x \, du + b_y \, dv + b_z \, dw) \, dV$$

Again the theorem requires that $d(U - V) = d\Pi = 0$.

6.5   FINITE ELEMENT APPLICATIONS

The solution of the tapered bar problem in section 5.3 typifies the displacement method in finite element analysis in which the displacements within each element are approximated in terms of their nodal point values, using suitable shape functions. The equations expressing equilibrium of the system of elements under the action of the actual external forces are then derived. The principle of virtual work provides a powerful tool for generalising this method by enabling the equilibrium equations to be derived indirectly.

We examine first the application of the principle to a single element of surface area S_e and volume V_e. The following sets of generalised forces and displacements are used.

(1) Equilibrium forces: the actual surface tractions $[t]$ and body forces $[b]$, together with the equilibrating stresses $[\sigma]$.

(2) Compatible displacements: the approximate displacements $[u]$, which are given in terms of the nodal values $[\delta^e]$ and shape functions $[N]$ as

$$[u] = [N][\delta^e]$$

The corresponding strain components in the element are

$$[e] = [B][\delta^e]$$

The requirement for the displacements to be compatible is met by choosing shape functions $[N]$ which are *continuous* within the element.

Using these sets of forces and displacements, together with the stress–strain law

$$[\sigma] = [D][e]$$

we can write the virtual work equation (6.9) as

$$\int_{S_e} [\delta^e]^T [N]^T [t] \, dS + \int_{V_e} [\delta^e]^T [N]^T [b] \, dV$$

$$= \int_{V_e} [\delta^e]^T [B]^T [D][B][\delta^e] \, dV$$

This reduces to

$$[F^e] = [K^e][\delta^e]$$

where the element stiffness matrix is given by

$$[K^e] = \int_{V_e} [B]^T [D][B] \, dV \tag{6.10}$$

and the column vector of equivalent nodal point forces is

$$[F^e] = \int_{S_e} [N]^T [t] \, dS + \int_{V_e} [N]^T [b] \, dV \tag{6.11}$$

If, in addition to body and surface forces, the solid is acted on by point loads, it is usual to arrange for there to be nodes placed at these points. There will then be an additional column vector $[P]$ on the right-hand side of equation (6.11) to account for these loads. Where a point load acts on an element at a location other than a node, the contribution to $[F^e]$ is found from the surface integral in (6.11) with $[t] \, dS = [P]$ and $[N]$ evaluated at the point of load application.

To apply the virtual work balance to a system of solid elements, we need to evaluate the area and volume integrals in (6.9). By writing these integrals as summations of the contributions made by each element, we obtain, once again, the algorithm (5.9) for the system stiffness matrix.

6.6 ENERGY BOUNDS IN THE DISPLACEMENT METHOD

In view of the equivalence between the virtual work balance and the theorem of minimum potential energy for an elastic solid, it follows that use of equations (6.10) and (6.11) leads to a finite element formulation which seeks to minimise the total potential energy Π of the system, within the constraints imposed by the choice of displacement functions $[u]$. Since the exact solution for the displacements gives an absolute minimum value for Π it follows that an approximate solution, by the displacement method, will always provide an *upper bound* on Π.

For a *linear* elastic solid the strain energy is equal to the work done by the external forces where

$$U = \sum_{n=1}^{n_n} \tfrac{1}{2} F_n \delta_n = \tfrac{1}{2} V$$

giving

$$\Pi = U - V = -U$$

Since the finite element solution for the displacements overestimates Π, the strain energy will be *underestimated*. Whilst this implies that the finite element model will be overstiff in an average sense, it does not necessarily follow that the displacement at any particular point is less than the exact value.

Consider the special case where a single point load W acts on a linear elastic structure, as in the example in section 5.4, and the approximate displacement at this point is ΔL. The strain energy is given by $U = \tfrac{1}{2} W \Delta L$ and, with U underestimated, it follows that ΔL is also underestimated, as confirmed by the results in Table 5.2.

The classical *Rayleigh–Ritz method* of structural analysis is based on the minimisation of potential energy where the displacements are approximated by single functions which span the entire structure. The finite element displacement method may be viewed as a form of the Rayleigh–Ritz method in which the displacements are approximated in a *piecewise* fashion by functions, each of which is defined over only a portion of the structure.

6.7 CHOICE OF DISPLACEMENT FUNCTIONS

For a finite element formulation to be acceptable the numerical solution must converge to the exact solution as the fineness of the discretisation is increased. In the displacement method, convergence is ensured if the following constraints are placed on the choice of the displacement functions $[u]$.

(1) They must define a displacement field which is continuous not only within each element but also between elements. The displacement method is based on the application of the virtual work principle both to individual elements and to the complete system of elements, and the applicability of the principle is restricted to sets of equilibrium forces and compatible displacements. Use of polynomial functions for $[u]$, as in (5.11) for the bar element, ensures continuity within

each element and at nodes common to more than one element. In the case of two- and three-dimensional problems, it is necessary to ensure, in addition, that continuity is maintained at every point along the interfaces between adjacent elements. For two-dimensional displacement fields the terms to be included in complete polynomials of various orders in x and y are given by *Pascal's triangle* (see Fig. 6.2).

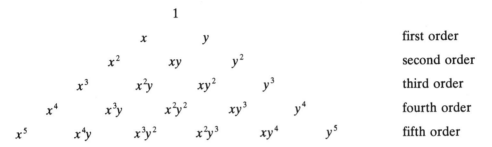

Fig. 6.2—Pascal's triangle for selecting complete polynomials of various orders.

(2) They must be capable of representing a constant strain state. This requirement follows from the fact that, as the elements are made progressively smaller, it is expected that this state will be approached within each element.
(3) They must be capable of representing rigid-body displacements. If an element has nodal point displacements corresponding to rigid-body rotation or translation, the choice of $[u]$ must ensure that the corresponding element strains $[e]$ are zero.

Elements whose displacement fields satisfy the compatibility condition (1) above are called *conforming*. Convergence is not restricted, however, only to such elements but is also possible, although not guaranteed, for certain *non-conforming* elements. These are elements which violate the inter-element compatibility condition by exhibiting overlapping and separation along their common edges (see Fig. 6.3). In this book, we restrict our attention to conforming elements.

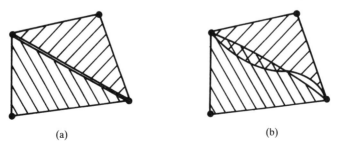

Fig. 6.3—Deformed interface between (a) conforming and (b) non-conforming elements.

D

PROBLEMS

6.1　By applying the principle of virtual work to an elastic, axially loaded two-node bar element of cross-section area $A = A(x)$ and length l, show that

$$[K^e] = \frac{EV}{l^2}\begin{bmatrix} 1 & -1 \\ -1 & 1 \end{bmatrix}$$

where $V = \int A\,dx$, is the volume of the element. For an element with a small uniform taper, show that $V = Al$, where A is the cross-section area of the element halfway along its length.

6.2　A linear elastic bar of uniform circular cross-section diameter D carries a shear stress $\tau = kx$ on its outer surface, where x is the axial coordinate and k is a constant. Using virtual work, show that the equivalent nodal point forces representing this loading for a two-node bar element are

$$[F^e] = \tfrac{1}{6}\pi Dl[2\tau_i + \tau_j \qquad \tau_i + 2\tau_j]^{\mathrm{T}}$$

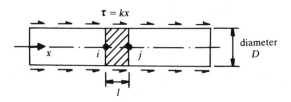

6.3　Using virtual work, show that the equivalent nodal point forces for a two-node bar element carrying a concentrated axial load P at $x = l/2$ are

$$[F^e] = \tfrac{1}{2}[P \qquad P]^{\mathrm{T}}$$

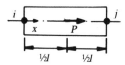

6.4　Show that the stiffness matrix for an axially loaded elastic bar element of uniform cross-section area A and with the three nodal points i, j and k is given by

$$[K^e] = \frac{EA}{l}\begin{bmatrix} 7 & -8 & 1 \\ -8 & 16 & -8 \\ 1 & -8 & 7 \end{bmatrix}$$

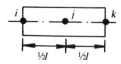

6.5   If the three-node elastic element in the previous problem hangs vertically under its own weight, show that the equivalent nodal point forces corresponding to the effect of gravity are

$$[F^e] = \tfrac{1}{6}\rho g A l [1 \quad 4 \quad 1]^{\mathrm{T}}$$

Chapter 7

A Computer Program for Two-dimensional Finite Element Stress Analysis

7.1 INTRODUCTION

In this chapter, we describe a computer program FIESTA2 for the finite element stress analysis of two-dimensional elastic problems. The program is a more advanced version of FIESTA1 (see Chapter 5) and uses algorithms for assembling and solving the equilibrium equations which are applicable to elements having any number of degrees of freedom. A complete listing of the source code is given in Appendix A3.

The element library contains elements for the analysis of frames, plane and axisymmetric solids and torsional problems (see Fig. 7.1). The number NODN of degrees of freedom per node, the number NONE of nodes per element, the number NOCO of coordinates required to define the element geometry, the number NOST of element stresses and the number NOPROP of material properties are defined for each type of element by means of DATA statements in subroutine SPEC. The formulation of the equilibrium equations for each type of element will be given in later chapters, together with a range of applications.

7.2 DATA INPUT

The complete specification of a finite element model requires the input of data to define the mesh geometry, the material properties of the elements and the boundary conditions. This information is supplied by the user in a format-free data file (see Fig. 7.2) read on unit 5. TITLE is the problem title which may contain up to 80 characters enclosed in apostrophes and IETYPE specifies the element type (see Fig. 7.1). By putting MFLAG = 0, we specify that the mesh data is to be read from the file. (The alternative is to put MFLAG = 1 and to make use of an automatic mesh generation facility to be described later.)

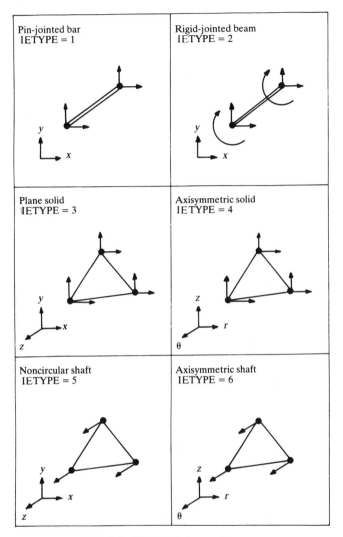

Fig. 7.1—FIESTA2 element library.

7.2.1 Mesh data

The following data defining the geometry of the mesh are read in subroutine RMESH: the number NONS of nodal points in the system, the coordinates XY(NS,1), XY(NS,2) of each node, the number NOE of elements in the system, and finally for each element the numbering IJ(IE,NE) of each node NE.

7.2.2 Material data

The material properties for the elements are read in subroutine PROPS. NOSET is the number of sets of properties to be defined, NOPROP is the number of properties per set, and the array EDAT stores the properties themselves. If

	TITLE IETYPE MFLAG		
Mesh data	MFLAG=0	NOSUB	MFLAG=1
	NONS XY(1,1),XY(1,2) . . . XY(NONS,1),XY(NONS,2) NOE IJ(1,1), . . IJ(1,NONE) . . . IJ(NOE,1), . . IJ(NOE,NONE)	NONR,NOES IFLAG(1), . . IFLAG(NOES) XYSUB(1,1),XYSUB(1,2) . . . XYSUB(8,1),XYSUB(8,2) . . for each subregion .	
Material data	NOSET EDAT(1,1), . . EDAT(1,NOPROP) . . EDAT(NOSET,1), . . EDAT(NOSET,NOPROP) MAT(1), . . MAT(NOE)		
Boundary conditions	NOPD IPD(1,1),IPD(1,2),PD(1) . . IPD(NOPD,1)IPD(NOPD,2),PD(NOPD) NOPS IPS(1,1), . . IPS(1,3),PS(1,1), . . PS(1,4) . . IPS(NOPS,1), . . IPS(NOPS,3),PS(NOPS,1), . . PS(NOPS,4) NOPF IPF(1,1),IPF(1,2),PF(1) . . IPF(NOPF,1),IPF(NOPF,2),PF(NOPF)		

Fig. 7.2—FIESTA2 data file for problem specification.

NOSET = 1, the same set MAT(IE) = 1 is assigned to each element IE without further data input by the user. Otherwise the set number is read from the data file for each element in turn.

7.2.3 Boundary conditions
Data defining any prescribed displacements, stresses or forces on the mesh boundary is read in subroutine BOUND. NOPD, NOPS and NOPF store the numbers of each type of boundary condition.

For each displacement boundary condition IBC the array elements IPD(IBC,1) and IPD(IBC,2) store the node and degree of freedom number, respectively, to

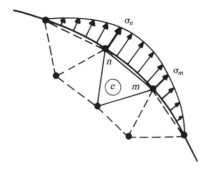

Fig. 7.3—Segment of mesh boundary carrying prescribed stresses.

which the boundary condition relates. PD(IBC) stores the prescribed value of the displacement. In the case of prescribed forces the arrays IPF and PF perform the equivalent tasks.

FIESTA2 allows *linearly varying* stresses to be prescribed over segments of the mesh boundary whose end points coincide with nodes (see Fig. 7.3). Each segment is defined by the boundary edges m, n of a series of elements e. IPS(IBC,1) identifies the element e, and IPS(IBC,2), IPS(IBC,3) the nodes m, n for each boundary condition IBC. PS(IBC,1), PS(IBC,2) store the prescribed values of the normal and shear stresses σ_m, τ_m at node m, and PS(IBC,3), PS(IBC,4) their prescribed values σ_n, τ_n at node n.

During the assembly of the system stiffness the equivalent set of nodal point forces $[F^e]$ for any element carrying prescribed stresses is stored in the array FE. It is therefore convenient to identify which stress boundary condition, if any, is associated with a particular element IE. This is achieved using an array JPS where JPS(IE) is either zero, where no stresses are prescribed, or else equal to the number of the boundary condition.

7.3 AUTOMATIC MESH GENERATION

For practical finite element meshes, which often contain many thousands of nodes and elements, the task of preparing the data is labour intensive and prone to human error. It is therefore desirable, where possible, to use the computer to *generate* the problem specification, and to keep the size of the data file to be prepared by the user to a minimum. Commercial programs generally have extensive data generation facilities, whose versatility and ease of use play an important role in determining their popularity with users.

FIESTA2 has a facility for generating meshes consisting of three-node triangular elements, of which there are several in the element library. To select this facility, we specify MFLAG=1 in the data file (see Fig. 7.2), in which case subroutine GMESH is called in place of RMESH. First a uniform square mesh is generated and then, using simple *coordinate mappings* defined in the data file, the mesh is fitted to the actual geometry of the problem. In this way, quite complex geometries can be modelled with relative ease and minimal data input.

7.3.1   Uniform square mesh

Using a local coordinate system (r, s) with its origin at the centre of a square of side length two units (see Fig. 7.4), we base the mesh on a uniform grid of n_r by n_s nodal points. The grid has step sizes Δr and Δs in the r and s directions, respectively, and a typical triangular element e has corner nodes i, j and k.

The mesh is generated in subroutine SQUARE where the variables NONR, NOES, DR, DS and S are used to store the values n_r, $n_s - 1$, Δr, Δs and s, respectively, and an array IFLAG stores flags for each of the NOES rows of elements in the s direction. To generate a uniform mesh we specify IFLAG(IR) = 0 for each row IR. The nodal point coordinates (r, s), generated for one line of nodes at a time in subroutine NLINE, are stored in the usual array XY. Array IJ stores the nodal point numbers i, j, k for each element. These are generated for one row of elements at a time in subroutine EROW using the repeating unit IUNIT (see Fig. 7.5) comprising the two elements IE1 and IE2.

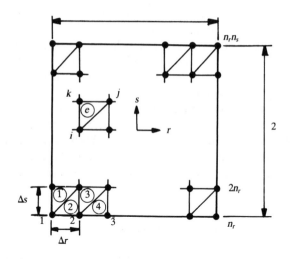

Fig. 7.4—Uniform square mesh.

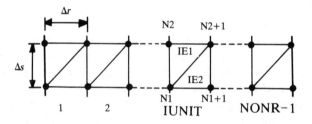

Fig. 7.5—Normal row of elements generated by subroutine EROW if IFLAG(IR) = 0.

7.3.2 Graded square mesh

The size of elements can be doubled or halved as successive rows are generated (see Fig. 7.6), by introducing one or more *transitional rows* of elements. These are generated by calling in place of subroutine EROW either EROW1 or EROW2. This is achieved by putting IFLAG(IR)=1 or IFLAG(IR)=2, respectively, for row IR. In either case, elements are generated using the repeating unit IUNIT (see Fig. 7.7) containing the three elements IE1, IE2 and IE3.

To avoid the need to calculate the initial value of the step size Δs, which will ensure that the last line of nodes lies exactly along the top edge of the square, the s coordinate of each node is scaled, if necessary, after the mesh generation is complete.

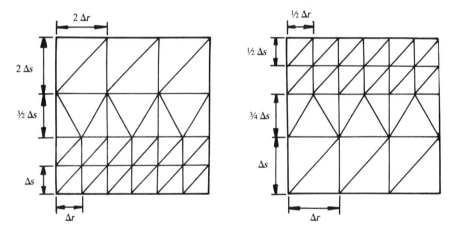

Fig. 7.6—Examples of graded square meshes generated using transitional rows of elements.

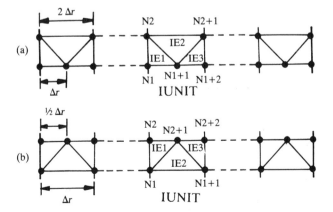

Fig. 7.7—Transitional rows of elements generated using (a) subroutine EROW1 to double element size if IFLAG(IR) = 1 and (b) subroutine EROW2 to halve element size if IFLAG(IR) = 2.

7.3.3 Coordinate mappings

If we apply *coordinate mappings* of the form

$$x = f(r, s) \quad \text{and} \quad y = g(r, s)$$

to the square mesh, it is possible to model complex quadrilateral regions by a suitable choice of the *mapping functions* f and g. The simplest mappings are the complete linear polynomials in r and s given by Pascal's triangle (see Fig. 6.2) as

$$x = C_1 + C_2 r + C_3 s + C_4 rs$$
$$y = D_1 + D_2 r + D_3 s + D_4 rs \tag{7.1}$$

and can be used to generate a *straight-sided* quadrilateral of arbitrary shape (see Fig. 7.8). The polynomial coefficients C_1, D_1, ... which define the mappings are found by matching f and g with the coordinates x and y at each of the four corners of the quadrilateral to give

$$C_1 + C_2 + C_3 + C_4 = x_1$$
$$C_1 - C_2 + C_3 - C_4 = x_2$$
$$C_1 - C_2 - C_3 + C_4 = x_3 \tag{7.2}$$
$$C_1 + C_2 - C_3 - C_4 = x_4$$

and a similar set of equations for the coefficients D_1, D_2, ... After solving equations (7.2), we can write (7.1) as

$$x = \sum_{i=1}^{4} N_i(r, s) x_i$$
$$y = \sum_{i=1}^{4} N_i(r, s) y_i \tag{7.3}$$

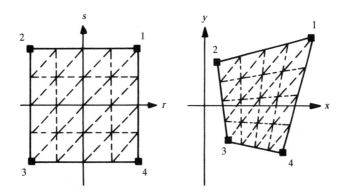

Fig. 7.8—Use of linear mapping functions.

where the *shape functions* N_i are given by

$$N_1 = \tfrac{1}{4}(1 + r)(1 + s), \qquad N_2 = \tfrac{1}{4}(1 - r)(1 + s)$$
$$N_3 = \tfrac{1}{4}(1 - r)(1 - s), \qquad N_4 = \tfrac{1}{4}(1 + r)(1 - s)$$

The complete *quadratic* polynomial mapping functions

$$x = C_1 + C_2 r + C_3 s + C_4 rs + C_5 r^2 + C_6 s^2 + C_7 rs^2 + C_8 r^2 s$$
$$y = D_1 + D_2 r + D_3 s + D_4 rs + D_5 r^2 + D_6 s^2 + D_7 rs^2 + D_8 r^2 s \tag{7.4}$$

are more versatile and their use enables a *curvilinear* quadrilateral region to be modelled (see Fig. 7.9). In addition to specifying the coordinates of the four corners, it is now necessary to specify the coordinates of one point on each of the four sides of the quadrilateral. By matching f and g with the (x, y) coordinates at each of the total of eight points we arrive at the following mappings:

$$x = \sum_{i=1}^{8} N_i(r, s) x_i$$
$$y = \sum_{i=1}^{8} N_i(r, s) y_i \tag{7.5}$$

where

$$N_i = \tfrac{1}{4}(1 + rr_i)(1 + ss_i)(rr_i + ss_i - 1) \qquad \text{for } i = 1, 3, 5, 8$$
$$= \tfrac{1}{2}(1 - r^2)(1 + ss_i) \qquad \text{for } i = 2, 6$$
$$= \tfrac{1}{2}(1 - s^2)(1 + rr_i) \qquad \text{for } i = 4, 8$$

If a side point does not lie on a straight line joining the adjacent corner points of the quadrilateral, then the side is described by a *parabola* passing through the three points.

For a uniform mesh the side points are positioned midway between the adjacent corners. The mesh is easily graded by moving the origin of the (r, s) coordinates

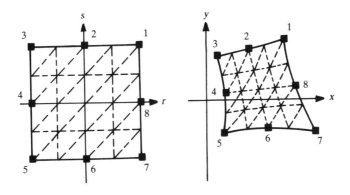

Fig. 7.9—Use of quadratic mapping functions.

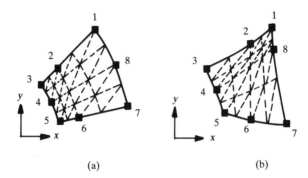

Fig. 7.10—Examples of curvilinear quadrilateral meshes graded (a) towards the side 3, 4, 5 and (b) towards the corner 1.

towards that side or corner of the mesh where smaller elements are required. This is achieved by shifting the side points accordingly (see Fig. 7.10).

Quadratic coordinate mappings (7.5) are performed in subroutine MAP. The coordinates of the eight points defining the curvilinear region are read from the data file and stored in the array XYSUB, whilst the shape functions SF are calculated in subroutine SHAPE.

7.3.4   Irregularly shaped meshes
Geometries of irregular shape are first divided up into a convenient number of curvilinear quadrilateral *subregions*. Meshes are then generated for each subregion in turn, using the techniques described above, and subsequently combined to form a single mesh. By this means, it is possible to model quite complex geometries with relative ease.

FIESTA2 generates meshes for geometries which can be modelled using a single chain of subregions in which each one is interfaced with not more than two others. The variable NOSUB, which stores the number of subregions to be generated, is read from the data file in subroutine GMESH. Some examples of the meshes that can be generated in this way are shown in Fig. 7.11.

7.4   FORMULATION OF THE EQUILIBRIUM EQUATIONS
We come now to that part of FIESTA2 in which the equilibrium equations for the system of elements are assembled and the boundary conditions satisfied.

7.4.1   Removal of constrained degrees of freedom
A simple method for imposing displacement boundary conditions, in which one or more displacements are specified, was given in section 5.3.4. For each boundary condition the relevant equilibrium equation is modified to enforce the condition.

Where a significant number of zero displacements are prescribed, it is usually more efficient to adopt a different approach in which the corresponding equations are deleted from the final set to be solved. In this way the boundary condition $\delta_i = 0$ is implemented by deleting row i and column i of $[K]$ and row i of $[F]$ thus:

$$
\begin{bmatrix}
K_{11} & K_{12} & \cdot & K_{1i} & \cdot & K_{1n} \\
K_{21} & K_{22} & \cdot & K_{2i} & \cdot & K_{2n} \\
\cdot & \cdot & & \cdot & & \cdot \\
K_{i1} & K_{i2} & \cdot & K_{ii} & \cdot & K_{in} \\
\cdot & \cdot & & \cdot & & \cdot \\
K_{n1} & K_{n2} & \cdot & K_{ni} & \cdot & K_{nn}
\end{bmatrix}
\begin{bmatrix}
\delta_1 \\
\delta_2 \\
\cdot \\
\delta_i \\
\cdot \\
\delta_n
\end{bmatrix}
=
\begin{bmatrix}
F_1 \\
F_2 \\
\cdot \\
F_i \\
\cdot \\
F_n
\end{bmatrix}
$$

where the equations remaining are referred to as the *active equations* of the system. The advantage of this method is that less memory is required for storing the equations and that the solution time for the problem is correspondingly reduced.

The additional book keeping required to identify the active equations, by means of an array IDA, is performed in subroutine ACTIVE. Array elements IDA(NS,IDN) are initially set to 1 for each degree of freedom IDN of each nodal point NS in the system. Those array elements corresponding to constrained degrees of freedom are then set to zero. Finally, the remaining nonzero elements of IDA are replaced with the corresponding active equation numbers IEQ. NOEQ stores the total number of active equations in the system.

After the solution for the nodal point displacements is complete, subroutine RENUM is called to renumber the displacements according to the original equation numbering scheme and to insert zeros for the constrained degrees of freedom.

7.4.2 Semi-bandwidth calculation

The one-dimensional elastic bar elements described in Chapter 5 were shown to give rise to system stiffnesses which are both symmetric and banded. Fortunately, $[K]$ retains these properties for two-dimensional elastic elements, thereby allowing substantial reductions to be made in both the computer storage and the solution time.

The semi-bandwidth of $[K]$ depends on the numbering scheme adopted for the nodal points and is determined by that element whose stiffness coefficients occupy positions in $[K]$ which are farthest from the leading diagonal (see Fig. 7.12). The semi-bandwidth IBW is calculated in subroutine BWIDTH where the column position ICS in the array SS for each element stiffness coefficient is calculated using the array IRC whose elements are conveniently defined in terms of the corresponding elements of the array IDA. If ICS is found to be greater than the previously calculated value of the semi-bandwidth, then IBW is updated accordingly.

7.4.3 Assembly of active equations

Subroutine ASSEM performs the assembly of the banded symmetric active equilibrium equations for a system of elements. Each element may have any

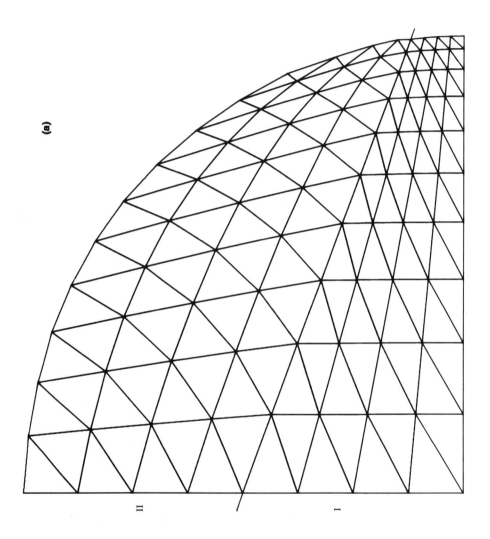

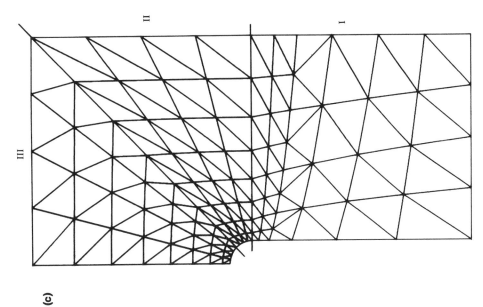

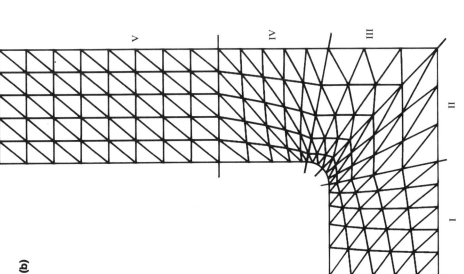

Fig. 7.11—Examples of irregularly shaped meshes generated by FIESTA2 using quadrilateral subregions.

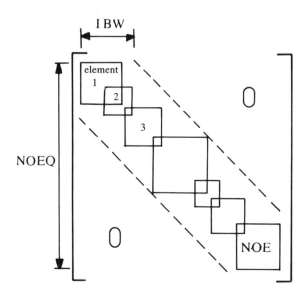

Fig. 7.12—Regions occupied by the coefficients of each element stiffness matrix within the system matrix.

number NONE of nodes and each node may have any number NODN of degrees of freedom. The subroutine is based on that given in Chapter 5 for use with one-dimensional elements.

Subroutine ELEMNT determines, according to the value of IETYPE, the subroutine employed to calculate the arrays SE and FE for each element in turn. Elements of the array IRC, which define the relationship between the row and column positions of stiffness coefficients in SE and SS, are defined in terms of the corresponding elements of IDA, as in subroutine BWIDTH. A zero element in the array IRC denotes a *nonactive equation* for which the assembly of SS is abandoned.

The nodal forces $[F]$ for the system of elements are found by summing the forces $[F^e]$ for each element. This is achieved by including the array FS in the assembly procedure for SS, where IRS is the system equation number equivalent to the element equation number IRE. Finally, any prescribed nodal point forces PF are added to FS using the array IDA to identify the active equation number for each boundary condition.

Any nonzero displacement boundary conditions are implemented using the method given in section 5.3.4.

7.5 SOLUTION OF THE EQUILIBRIUM EQUATIONS

The Gaussian elimination method described in section 5.3.5 for a tridiagonal set of linear equations can be generalised for a set of equations of arbitrary bandwidth. We examine first, however, the application of the method to an unbanded set of equations.

7.5.1 Unbanded equations
The forward reduction process for the set of equations

$$
\begin{bmatrix}
K_{11} & K_{12} & . & . & . & K_{1n} \\
K_{21} & K_{22} & . & . & . & K_{2n} \\
. & . & . & . & . & . \\
. & . & . & . & . & . \\
. & . & . & . & . & . \\
K_{n1} & K_{n2} & . & . & . & K_{nn}
\end{bmatrix}
\begin{bmatrix}
\delta_1 \\ \delta_2 \\ . \\ . \\ . \\ \delta_n
\end{bmatrix}
=
\begin{bmatrix}
F_1 \\ F_2 \\ . \\ . \\ . \\ F_n
\end{bmatrix}
\tag{7.6}
$$

begins with the elimination of δ_1 which will, in general, be present in all equations. It follows that the second equation through to equation n are affected, by contrast with a tridiagonal set where only the second equation is affected. The elimination of δ_2 extends from the third equation through to equation n, and so on. After the elimination of δ_i using equation i, the coefficients of the subsequent equations are

$$K'_{rc} = K_{rc} - K_{ic}R \tag{7.7}$$

$$F'_r = F_r - F_i R \tag{7.8}$$

where $R = K_{ri}/K_{ii}$ and both the row index r and the column index c range from $i + 1$ to n. The extent of reduction step i is indicated by the broken boxes in Fig. 7.13. Since only the coefficients on or above the leading diagonal are subsequently used during back substitution, it is not necessary actually to replace the coefficients of δ_i by zero, although this is, of course, implied by the definition of

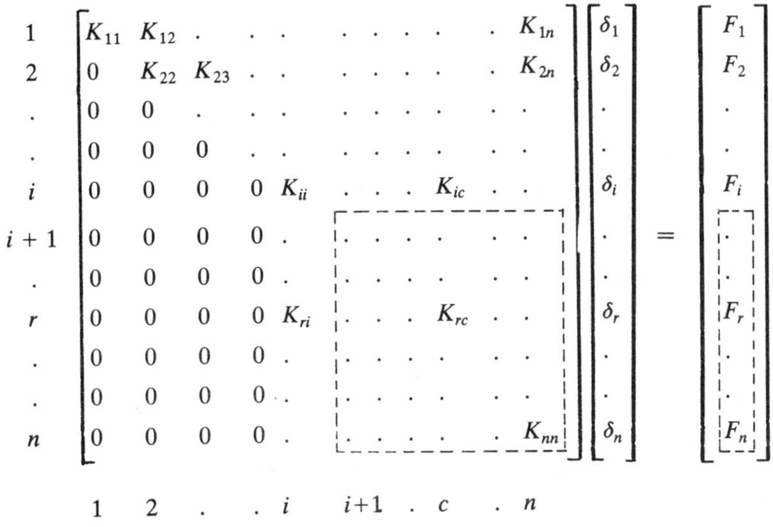

Fig. 7.13—Extent of the forward reduction step i for a set of unbanded equations.

the elimination ratio R. Forward reduction continues until $i = n - 1$, after which the equations are of the upper triangular form

$$
\begin{bmatrix}
K_{11} & K_{12} & . & . & . & . & K_{1n} \\
0 & K_{22} & K_{23} & . & . & . & K_{2n} \\
0 & 0 & . & . & . & . & . \\
0 & 0 & 0 & . & . & . & . \\
0 & 0 & 0 & 0 & . & . & . \\
0 & 0 & 0 & 0 & 0 & . & . \\
0 & 0 & 0 & 0 & 0 & 0 & K_{nn}
\end{bmatrix}
\begin{bmatrix}
\delta_1 \\
\delta_2 \\
. \\
. \\
. \\
\delta_{n-1} \\
\delta_n
\end{bmatrix}
=
\begin{bmatrix}
F_1 \\
F_2 \\
. \\
. \\
. \\
F_{n-1} \\
F_n
\end{bmatrix}
\tag{7.9}
$$

Back substitution commences with equation n from which we obtain

$$
\delta_n = \frac{F_n}{K_{nn}}
\tag{7.10}
$$

and, working back through equations $n - 1, n - 2, \ldots$, we reach equation r which yields the solution

$$
\delta_r = \frac{F_r - \sum\limits_{c=r+1}^{n} K_{rc}\delta_c}{K_{rr}}
\tag{7.11}
$$

and so on, until $r = 1$.

7.5.2 Banded symmetric equations

The solution of a set of n banded symmetric equations of semi-bandwidth b is performed in subroutine SOLVE where the matrix $[K]$ is stored as an n by b array SS, as shown in Fig. 7.14. The first column of SS contains the stiffness coefficients on the leading diagonal of $[K]$, and the lower right-hand corner has zeros for those coefficients which lie beyond column n in matrix $[K]$.

A forward reduction step does not destroy the symmetry of that region of coefficients in $[K]$ affected by the next step. It follows that K_{ir} may be substituted for K_{ri} in the calculation of the elimination ratio R. Equations (7.7)–(7.11) are re-expressed in terms of the equivalent elements of the array SS using the relationships in Table 5.1. The elements of SS affected by step i are indicated by the triangular broken box in Fig. 7.14. Those involved in the reduction of equation r lie within the rectangular broken boxes. The diagonal decay test for ill-conditioning, described in section 5.5, is performed after each reduction step.

Each forward reduction step i need now only extend, at most, over a block of $b - 1$ equations. Forward reduction beyond row n of matrix $[K]$ is prevented by avoiding the triangular region of zeros in the bottom right-hand corner of the array

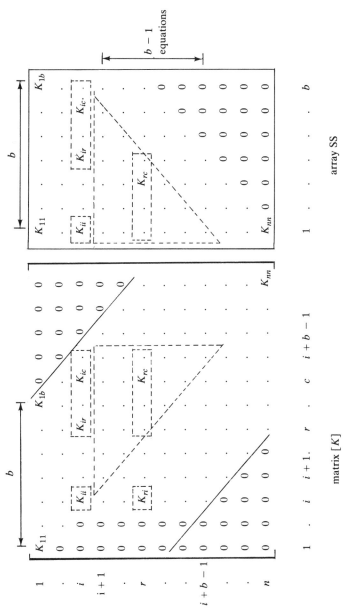

Fig. 7.14—Storage scheme for a symmetric banded matrix $[K]$ showing the coefficients involved in the forward reduction step i.

SS. The reduction of equation r need only extend from column $c = r$ through to, at most, column $i + b - 1$ in matrix $[K]$. Back substitution for equation r need only extend from column r to, at most, column $r + b - 1$ in $[K]$.

In commercial programs the forward reduction of the nodal point forces $[F]$ is usually performed as a separate operation, after the reduction of $[K]$ is complete. This requires the value of the eliminaton ratio R for each step to be stored and facilitates the efficient analysis of *multiple-load cases* for a given structure. Forward reduction of $[K]$ is performed only once, whilst forward reduction of $[F]$ and back substitution are carried out for each load case in turn.

Despite the efficiency of the storage scheme for $[K]$ described above, it may not be possible to hold the entire array SS in the computer core memory for problems with a very large number of degrees of freedom. In Fig. 7.14 it is seen, however, that only a limited triangular region of stiffness coefficients need be available in the memory at any one time. This fact is exploited in the *block solver scheme* in which the complete set of equations is stored on disc in blocks. Each block contains b equations, and two blocks are held in memory at a time. When the upper block of equations has been reduced, it is transferred to disc, the lower block is shifted up in the memory and a new block read from disc. The importance of using a nodal point numbering scheme which minimises the semi-bandwidth of $[K]$ is obvious.

Probably the most efficient form of Gaussian elimination is provided by the *frontal solution scheme* in which the equations are assembled and reduced in a single-step operation. At no time is it necessary to have the complete set of assembled equations available. The sequence of operations is determined by the advance of a front of the unknowns through the finite element mesh. The width of this front is determined by the element-numbering scheme adopted.

7.6 CALCULATION OF ELEMENT STRESSES

When the solution for the nodal point displacmeements $[\delta]$ is complete, the stresses in each element are calculated. Each type of element has a subroutine for this purpose selected by subroutine STRESS according to the value of IETYPE.

To economise on computer memory, the array SS which holds the system stiffness during the assembly and solution of the equilibrium equations is thereafter re-used to store the element stresses. A total of NOST stress components are calculated for each element and SS(IE,IST) stores the component IST for element IE.

7.7 PROGRAM OUTPUT

All output is directed to a suitable device by way of unit 6. A complete listing of the problem specification is performed by subroutine LIST and should be thoroughly checked by the user for any errors or omissions. The absence of execution errors when the program is run should never be used as evidence that the problem has been correctly specified. It is quite possible for the data file to specify correctly a problem different to that intended by the user!

The solutions obtained for the nodal point displacements and element stresses are stored in arrays FS and SS, respectively, and are listed in subroutine OUTPUT. This output should be carefully examined for any obvious anomalies and, where possible, should be compared with the results obtained using a simplified analytical model of the problem.

Chapter 8

Finite Element Analysis of Elastic Frames

8.1 INTRODUCTION

In this chapter, we examine the application of the finite element method to the analysis of *planar elastic frames*. A planar or two-dimensional frame has all its members lying in a single plane and all applied loads act in this plane (see Fig. 8.1).

Two idealised models are commonly used, depending on the behaviour of the joints. In a *pin-jointed frame* the members are assumed to be connected to each other by frictionless pins which allow free rotation and cannot transmit bending moments. Provided that loads are applied only at the joints, it follows that the members will be subject only to axial forces. By contrast, in a *rigid-jointed frame* the ends of all members connected by a particular joint undergo the same rotation, and loads may be applied along the lengths of the members as well as at the joints. In addition to axial forces, members also carry shear forces and bending moments.

To model a plane frame with a system of finite elements the choice of subdivision is obvious; normally, it suffices to represent each member by a single element. For this reason the analysis of plane frames provides a useful introduction to the finite element analysis of two-dimensional problems for which, in general, the subdivision is not self-evident.

8.2 PIN-JOINTED FRAMES

Since individual members are subjected only to axial forces, their behaviour is modelled by the bar element described in section 5.3.

8.2.1 Finite element formulation

The elastic behaviour of a two-node bar element in a local axial coordinate system $\bar{x}$

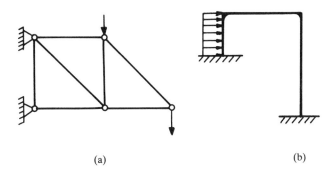

(a)   (b)

Fig. 8.1—(a) Pin-jointed and (b) rigid-jointed planar frames.

(see Fig. 8.2(a)) is given by (5.16) as

$$\begin{bmatrix} F_i^e \\ F_j^e \end{bmatrix} = \frac{EA}{l} \begin{bmatrix} 1 & -1 \\ -1 & 1 \end{bmatrix} \begin{bmatrix} \delta_i \\ \delta_j \end{bmatrix} \tag{8.1}$$

or

$$[\bar{F}^e] = [\bar{K}^e][\bar{\delta}^e]$$

The axis of the element is inclined at an angle α to the x direction of the global (x, y) coordinate system, as shown in Fig. 8.2(b). Before a system stiffness matrix can be assembled, it is necessary to transform (8.1) to this system. In global coordinates, each node has *two* degrees of freedom u and v, the components of displacement in the x and y directions, respectively. The corresponding force components are denoted by P^e and Q^e. The relationships between nodal forces and displacements in the local and global coordinate systems are

$$P_i^e = F_i^e \cos \alpha, \qquad Q_i^e = F_i^e \sin \alpha, \qquad P_j^e = F_j^e \cos \alpha, \qquad Q_j^e = F_j^e \sin \alpha$$

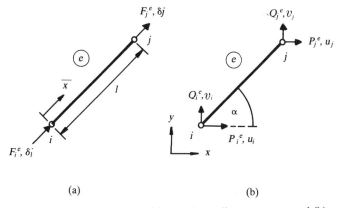

(a)   (b)

Fig. 8.2—Typical pin-jointed element in (a) a local coordinate system $\bar{x}$ and (b) a global coordinate system (x, y).

and

$$\delta_i = u_i \cos \alpha + v_i \sin \alpha, \qquad \delta_j = u_j \cos \alpha + v_i \sin \alpha$$

or in matrix notation

$$\begin{bmatrix} P_i^e \\ Q_i^e \\ P_j^e \\ Q_j^e \end{bmatrix} = \begin{bmatrix} c & 0 \\ s & 0 \\ 0 & c \\ 0 & s \end{bmatrix} \begin{bmatrix} F_i^e \\ F_j^e \end{bmatrix}, \qquad \begin{bmatrix} \delta_i \\ \delta_j \end{bmatrix} = \begin{bmatrix} c & s & 0 & 0 \\ 0 & 0 & c & s \end{bmatrix} \begin{bmatrix} u_i \\ v_i \\ u_j \\ v_j \end{bmatrix}$$

or simply

$$[F^e] = [T]^{\mathrm{T}}[\bar{F}^e] \qquad \text{and} \qquad [\bar{\delta}^e] = [T][\delta^e] \tag{8.2}$$

where $c = \cos \alpha$, $s = \sin \alpha$ and $[T]$ is a *transformation matrix*. We now combine equations (8.1) and (8.2) to give

$$[F^e] = [T]^{\mathrm{T}}[\bar{F}^e] = [T]^{\mathrm{T}}[\bar{K}^e][\bar{\delta}^e]$$
$$= [T]^{\mathrm{T}}[\bar{K}^e][T][\delta^e]$$
$$= [K^e][\delta^e]$$

where the element stiffness matrix in the global coordinate system is

$$[K^e] = [T]^{\mathrm{T}}[\bar{K}^e][T] \tag{8.3}$$

$$= \frac{EA}{l} \begin{bmatrix} c^2 & cs & -c^2 & -cs \\ cs & s^2 & -cs & -s^2 \\ -c^2 & -cs & c^2 & cs \\ -cs & -s^2 & cs & s^2 \end{bmatrix} \tag{8.4}$$

The axial stress in an element e is given in terms of the axial displacements $[\bar{\delta}^e]$ by (5.26) as

$$[\sigma] = \frac{E}{l}[-1 \quad 1][\bar{\delta}^e]$$

Substituting for $[\bar{\delta}^e]$ from (8.2), we can write this as

$$[\sigma] = \frac{E}{l}[-1 \quad 1][T][\delta^e]$$
$$= \frac{E}{l}[-c \quad -s \quad c \quad s][\delta^e] \tag{8.5}$$

8.2.2 Use of FIESTA2
To select the appropriate subroutine ELEM1 from the element library of FIESTA2, we put IETYPE=1 in the data file specifying the finite element model

(see Fig. 7.2). The automatic mesh generation facility is not available for this element.

Subroutine ELEM1 calculates the element stiffness array SE using (8.4) and the array of element stresses SS is calculated in subroutine STR1 using (8.5).

The material data for the problem are specified in the manner described in section 7.2.2. The NOPROP=2 properties for each property set ISET are stored in the array EDAT as follows:

$$\text{EDAT(ISET,1)} = E, \qquad \text{EDAT(ISET,2)} = A$$

Force and displacement boundary conditions are specified as explained in section 7.2.3, and care should be taken to ensure that a sufficient number of displacements are prescribed; otherwise, $[K]$ will be a singular matrix. In this case, FIESTA2 prints the error message 'ILL-CONDITIONED EQUATIONS' and execution of the program is terminated. It is not, of course, possible to specify stress boundary conditions for this element.

8.2.3 Analysis of a bridge truss
The simple bridge truss shown in Fig. 8.3 consists of nine pin-jointed elastic members each with the same elastic modulus E. Joint 1 is attached to a rigid support whilst joint 3 is free to move in the horizontal direction. Vertical loads $2W$ and $3W$ are applied at joints 5 and 6, respectively. Members 1 to 6 have a cross-section area $2 \times 10^{-4} L^2$ and members 7, 8 and 9 have a cross-section area $10^{-4} L^2$. A finite element analysis of the problem using nine elements involves the solution of nine active equations. The results obtained for the deflected shape of the truss and the axial stresses in the members are shown in Fig. 8.4. In practice the joints will inevitably offer some frictional resistance to rotation and therefore the actual joint deflections will be less than those calculated here.

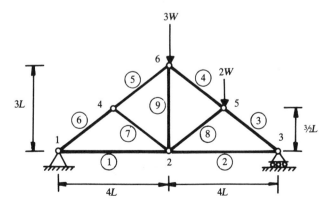

Fig. 8.3—Bridge truss problem.

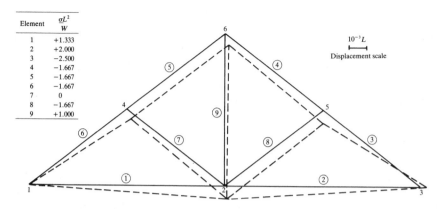

Element	$\dfrac{\sigma L^2}{W}$
1	+1.333
2	+2.000
3	−2.500
4	−1.667
5	−1.667
6	−1.667
7	0
8	−1.667
9	+1.000

Fig. 8.4—Deflected shape and stresses for the bridge truss problem.

8.3 RIGID-JOINTED FRAMES

Members of a rigid-jointed frame carry both axial and bending loads. According to the principle of superposition (see section 4.5) the combined effect of these two types of load can be found by summing their individual effects. The response of a member to axial loads is modelled by the bar element described in section 5.3. To model the effect of bending loads, we employ a beam element.

8.3.1 Beam element

The simplest beam element (see Fig. 8.5) has two nodes i and j and is of uniform cross-section area. The derivation of $[K^e]$ is based on elementary beam theory for the pure symmetric bending of an elastic beam.

The generalised nodal point displacements are here the lateral deflection w and the rotation $\theta = dw/dx$, whilst the corresponding generalised nodal point forces are the shear force V and the bending moment M. Thus,

$$[\delta^e] = [w_i \quad \theta_i \quad w_j \quad \theta_j]^T, \qquad [F^e] = [V_i \quad M_i \quad V_j \quad M_j]^T$$

The appropriate choice of displacement function for an element with four degrees of freedom is the *cubic* polynomial

$$[u] = w(x) = C_0 + C_1 x + C_2 x^2 + C_3 x^3 \tag{8.6}$$

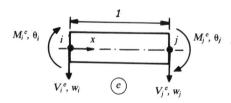

Fig. 8.5—Typical beam element.

The values of $w(x)$ and $\theta(x) = dw/dx$ at the nodes will match the corresponding nodal displacements if

$$
\begin{bmatrix} w_i \\ \theta_i \\ w_j \\ \theta_j \end{bmatrix} = \begin{bmatrix} 1 & 0 & 0 & 0 \\ 0 & 1 & 0 & 0 \\ 1 & l & l^2 & l^3 \\ 0 & 1 & 2l & 3l^2 \end{bmatrix} \begin{bmatrix} C_0 \\ C_1 \\ C_2 \\ C_3 \end{bmatrix}
$$

which yields the following solution for the coefficients:

$$
\begin{bmatrix} C_0 \\ C_1 \\ C_2 \\ C_3 \end{bmatrix} = \frac{1}{l^3} \begin{bmatrix} l^3 & 0 & 0 & 0 \\ 0 & l^3 & 0 & 0 \\ -3l & -2l^2 & 3l & -l^2 \\ 2 & l & -2 & l \end{bmatrix} \begin{bmatrix} w_i \\ \theta_i \\ w_j \\ \theta_j \end{bmatrix}
\tag{8.7}
$$

We can now write the displacement function as

$$
[u] = [N_1 \ \ N_2 \ \ N_3 \ \ N_4][\delta^e] = [N][\delta^e]
\tag{8.8}
$$

where the cubic shape functions are

$$
N_1 = 1 - 3X^2 + 2X^3, \qquad N_2 = l(X - 2X^2 + X^3)
$$
$$
N_3 = 3X^2 - 2X^3, \qquad N_4 = l(-X^2 + X^3)
$$

with $X = x/l$.

The generalised strain for a beam element corresponds to the *curvature*

$$
[e] = \frac{-d^2 w}{dx^2}
$$

$$
= \frac{-1}{l^2} \frac{d^2}{dx^2} ([N_1 \ \ N_2 \ \ N_3 \ \ N_4][\delta^e])
\tag{8.9}
$$

$$
= [B][\delta^e]
$$

where

$$
[B] = \frac{1}{l^2} [6 - 12X \quad l(4 - 6X) \quad -6 + 12X \quad l(2 - 6X)]
$$

and the generalised stress, in the absence of shear deformation, is the bending moment M given by

$$[\sigma] = M = -EI\frac{d^2w}{dx^2} = [D][e] \tag{8.10}$$

where $[D] = EI$ is the *flexural rigidity* of the element.

By substituting for $[B]$ from (8.9) and $[D]$ from (8.10) into (6.10), we obtain the element stiffness matrix as

$$[K^e] = \int_{V_e^1} [B]^T[D][B] \, dV$$

$$= l\int_0^1 [B]^T[D][B] \, dX$$

$$= \frac{EI}{l^3}\begin{bmatrix} 12 & 6l & -12 & 6l \\ & 4l^2 & -6l & 2l^2 \\ & & 12 & -6l \\ & \text{symmetric} & & 4l^2 \end{bmatrix} \tag{8.11}$$

The equivalent set of nodal point forces corresponding to a linearly varying distributed load (see Fig. 8.6) of intensity

$$[t] = q = q_i(1 - X) + q_jX$$

is obtained by substituting for $[N]$ from (8.8) into (6.11) and integrating to give

$$[F^e] = \int_{S_e} [N]^T[t] \, dS$$

$$= \frac{l}{60}\begin{bmatrix} 21 & 9 \\ 3l & 2l \\ 9 & 21 \\ -2l & -3l \end{bmatrix}\begin{bmatrix} q_i \\ q_j \end{bmatrix} \tag{8.12}$$

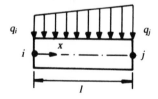

Fig. 8.6—Beam element carrying a linearly varying distributed load.

These forces are to be added to any concentrated loads or bending moments
applied at the nodes.

8.3.2 Combined beam and bar element

The combined effect of axial and bending loads is represented by an element whose
stiffness matrix is obtained by combining $[K^e]$ for a bar element, given by (5.16),
with $[K^e]$ for the beam element, given by (8.11). The resulting equilibrium
equations in the local coordinate system $(\bar{x}, \bar{y})$ (see Fig. 8.7(a)) are

$$
\begin{bmatrix} F_i^e \\ V_i^e \\ M_i^e \\ F_j^e \\ V_j^e \\ M_j^e \end{bmatrix} =
\begin{bmatrix}
EA/l & 0 & 0 & -EA/l & 0 & 0 \\
 & 12EI/l^3 & 6EI/l^2 & 0 & -12EI/l^3 & 6EI/l^2 \\
 & & 4EI/l & 0 & -6EI/l^2 & 2EI/l \\
\text{symmetric} & & & EA/l & 0 & 0 \\
 & & & & 12EI/l^3 & -6EI/l^2 \\
 & & & & & 4EI/l
\end{bmatrix}
\begin{bmatrix} \delta_i \\ w_i \\ \theta_i \\ \delta_j \\ w_j \\ \theta_j \end{bmatrix}
$$

or

$$ [\bar{F}^e] = [\bar{K}^e][\bar{\delta}^e] \tag{8.13} $$

We now transform these equations to the global coordinate system (x, y) (see
Fig. 8.7(b)) in which the three nodal degrees of freedom are the displacements u, v
in the x, y directions and a rotation θ. The corresponding generalised forces P^e, Q^e

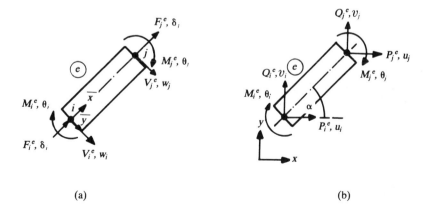

(a)   (b)

Fig. 8.7—Nodal forces and displacements for a combined bar and beam element in (a) local
coordinates and (b) global corrdinates.

and M^e are related to the forces in the local coordinate system by

$$
\begin{bmatrix} P_i^e \\ Q_i^e \\ M_i^e \\ P_j^e \\ Q_j^e \\ M_j^e \end{bmatrix} = \begin{bmatrix} c & s & 0 & 0 & 0 & 0 \\ s & -c & 0 & 0 & 0 & 0 \\ 0 & 0 & 1 & 0 & 0 & 0 \\ 0 & 0 & 0 & c & s & 0 \\ 0 & 0 & 0 & s & -c & 0 \\ 0 & 0 & 0 & 0 & 0 & 1 \end{bmatrix} \begin{bmatrix} \bar{F}_i^e \\ \bar{V}_i^e \\ \bar{M}_i^e \\ \bar{F}_j^e \\ \bar{V}_j^e \\ \bar{M}_j^e \end{bmatrix}
$$

or

$$[F^e] = [T]^\mathrm{T}[\bar{F}^e] = [T][\bar{F}^e] \tag{8.14}$$

and the relationship between the displacements is

$$[\bar{\delta}^e] = [T][\delta^e] \tag{8.15}$$

where $[T]$ is the transformation matrix. The element stiffness matrix in global coordinates, found using equation (8.3) with $[\bar{K}^e]$ defined in (8.13) and $[T]$ in (8.14), is given by

$$[K^e] = [T]^\mathrm{T}[\bar{K}^e][T]$$

$$
= \begin{bmatrix}
c^2e + s^2f & cs(e - f) & sg & -c^2e - s^2f & cs(-e + f) & sg \\
 & s^2e + c^2f & -cg & cs(-e + f) & -s^2e - c^2f & -cg \\
 & & h & -sg & cg & \tfrac{1}{2}h \\
 & \text{symmetric} & & c^2e + s^2f & cs(e - f) & -sg \\
 & & & & s^2e + c^2f & cg \\
 & & & & & h
\end{bmatrix}
\tag{8.16}
$$

where $e = EA/l$, $f = 12EI/l^3$, $g = 6EI/l^2$ and $h = 4EI/l$.

The equivalent set of nodal point forces, in global coordinates, for a linearly varying distributed load is found by substituting (8.12) into (8.14) to give

$$
\begin{bmatrix} P_i^e \\ Q_i^e \\ M_i^e \\ P_j^e \\ Q_j^e \\ M_j^e \end{bmatrix} = \frac{l}{60} \begin{bmatrix} 21s & 9s \\ -21c & -9c \\ 3l & 2l \\ 9s & 21s \\ -9c & -21c \\ -2l & -3l \end{bmatrix} \begin{bmatrix} q_i \\ q_j \end{bmatrix}
\tag{8.17}
$$

After the solution for the displacements is complete, we calculate the stress resultants in each element by expressing $[\bar{F}^e]$ in (8.13) in terms of the displacements $[\delta^e]$ using (8.15) to give

$$[\bar{F}^e] = [\bar{K}^e][\bar{\delta}^e] = [\bar{K}^e][T][\delta^e]$$

or

$$
\begin{bmatrix} F_i^e \\ V_i^e \\ M_i^e \\ F_j^e \\ V_j^e \\ M_j^{el} \end{bmatrix} =
\begin{bmatrix}
ce & se & 0 & -ce & -se & 0 \\
sf & -cf & g & -sf & cf & g \\
sg & -cg & h & -sg & cg & \frac{1}{2}h \\
-ce & -se & 0 & ce & se & 0 \\
-sf & cf & -g & sf & -cf & -g \\
sg & -cg & \frac{1}{2}h & -sg & cg & h
\end{bmatrix}
\begin{bmatrix} u_i \\ v_i \\ \theta_i \\ u_j \\ v_j \\ \theta_j \end{bmatrix}
\tag{8.18}
$$

8.3.3 Use of FIESTA2

To select the appropriate element subroutine ELEM2 in FIESTA2 we put IETYPE=2 in the data file specifying the problem (see Fig. 7.2). The automatic mesh generation facility is not available for this element.

Subroutine ELEM2 calculates the element stiffness array SE using (8.16), and the array FE of nodal point forces corresponding to a linearly varying distributed load q using (8.17). The array of element stress resultants SS is calculated in subroutine STR2 using (8.18).

The material data for the problem are specified in the manner described in section 7.2.2. The NOPROP=3 properties for each property set ISET are stored in the array EDAT as follows:

EDAT(ISET,1) = E,   EDAT(ISET,2) = A,   EDAT(ISET,3) = I

Force and displacement boundary conditions are specified in the manner described in section 7.2.3. Unless a sufficient number of displacements are prescribed, $[K]$ will be a singular matrix and execution of FIESTA2 will terminate with the error message 'ILL-CONDITIONED EQUATIONS'. The prescribed values for stress boundary condition IBC are stored in the array PS as follows:

PS(IBC,1) = q_i,   PS(IBC,2) = 0

PS(IBC,3) = q_j,   PS(IBC,4) = 0

8.3.4 Analysis of a two-storey building frame

The two-storey rigid-jointed building frame shown in Fig. 8.8 is subjected to horizontal loads $2W$ and W at joints 4 and 7, respectively. Members 4 and 5 carry uniformly distributed loads of intensities $4W/L$ and $2W/L$, respectively. All

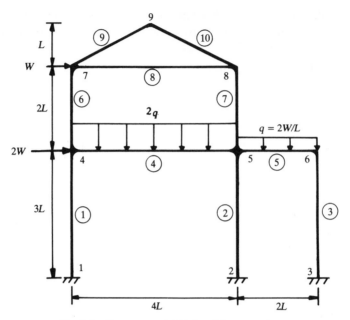

Fig. 8.8—Two-storey rigid-jointed frame problem.

members have the same elastic modulus E and all members have $A = 10^{-3}L^2$, $I = 10^{-5}L^4$ except 4 and 5 for which $A = 2 \times 10^{-3}L^2$, $I = 2 \times 10^{-5}L^4$. Fig. 8.9 shows the deflected shape of the frame obtained using a finite element model consisting of ten elements and a total of 18 active equations.

Special care should be exercised in the application of FIESTA2 to this type of problem which can result in a set of ill-conditioned equations (see section 5.5) when the axial stiffnesses EA/l are several orders of magnitude greater than the bending stiffnesses EI/l^3. Although the members undergo very small axial strains, it is not permissible to remove their axial stiffnesses from the formulation of $[K^e]$, without at the same time preventing their change in length by imposing suitable displacement boundary conditions.

8.3.5 Curved beams

A curved beam can be analysed in an approximate manner by treating it as an assemblage of straight beam elements. The analysis procedes exactly as in the case of a rigid-jointed frame. As the number of elements is increased, the geometry of the mesh approaches that of the curved beam.

Consider the *thin* ring of mean radius a shown in Fig. 8.10(a). The ring is loaded by a pair of equal and opposite diametral forces W and has a uniform square cross-section of side length $a/10$. The exact solution for the distribution of the bending moment in the ring is given by

$$M = \frac{1}{2} Wa \left(\frac{2}{\pi} - \sin \theta \right) \tag{8.19}$$

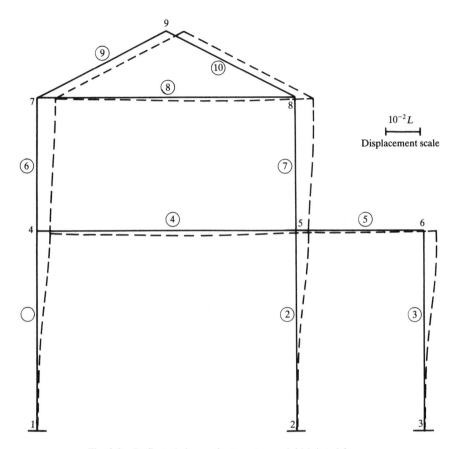

Fig. 8.9—Deflected shape of a two-storey rigid-jointed frame.

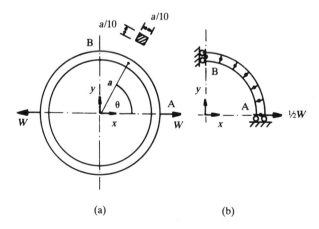

(a)   (b)

Fig. 8.10—(a) Thin ring loaded by diametral forces and (b) the finite element mesh for one quadrant of the ring.

and for the radial displacements at A and B by

$$u_A = \frac{Wa^3}{8EI}\left(\pi - \frac{8}{\pi}\right), \qquad v_B = -\frac{Wa^3}{4EI}\left(\frac{4}{\pi} - 1\right) \qquad (8.20)$$

Since the ring has two planes of geometric symmetry $x = 0$ and $y = 0$, it is only necessary to model a single quadrant AB using the finite element mesh shown in Fig. 8.10(b). It is apparent that the results obtained for the displacements u_A and v_B (see Table 8.1) converge to the exact solutions given by (8.20) as the number of

Table 8.1   Finite element results for radial displacements in a thin ring loaded by diametral forces

number of elements	$(u_b Ea/W) \times 10^3$	$(v_b Ea/W) \times 10^3$
3	8.496	−7.738
6	8.840	−8.058
9	8.910	−8.120
12	8.924	−8.142
Exact solution	8.927	−8.197

elements is increased. The bending moment distribution for the most refined mesh with 12 elements (see Fig. 8.11) is in very close agreement with the exact solution (8.19).

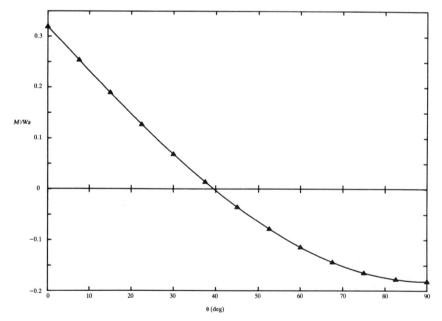

Fig. 8.11—Variation in the bending moment around a thin ring loaded by diametal forces: ▲, finite element results; ——, exact solution (8.19).

PROBLEMS

8.1   For a pin-jointed bar element show that the axial strain $[e]$ is related to the nodal displacements $[\delta^e]$ in the global (x, y) coordinate system by $[e] = [B][\delta^e]$ where

$$[B] = \frac{1}{l}[-c \quad -s \quad c \quad s]$$

Hence, verify equation (8.4) for the element stiffness matrix using virtual work.

8.2   An elastic cantilever carries a uniformly distributed load of intensity q over its length L. Use a single beam element to represent the problem and show that the solutions for the deflection and rotation at the free end are the exact values. Deduce the distribution of bending moment along the beam and compare it with the exact solution.

8.3   To include the effect of the weight of the members of a rigid-jointed frame, each element is assumed to have a weight ρ per unit length. Find the equivalent set of nodal forces $[F^e]$ in the global (x, y) coordinate system, where x is the horizontal coordinate.

(Answers: $\rho l[0 \quad -\frac{1}{2} \quad \frac{1}{12}l \cos \alpha \quad 0 \quad -\frac{1}{2} \quad -\frac{1}{12}l \cos \alpha]^{\mathrm{T}}$.)

8.4   Using virtual work, find the equivalent nodal point forces $[F^e]$ in the local coordinate system $\bar{x}$ for a beam element carrying a concentrated lateral load P at its midspan.

(Answer: $P[\frac{1}{2} \quad \frac{1}{8}l \quad \frac{1}{2} \quad -\frac{1}{8}l]^{\mathrm{T}}$.)

Chapter 9

Plane Elasticity Problems

9.1 INTRODUCTION

The plane theory of elasticity is used to formulate stress analysis problems for solids bounded by two parallel plane stress-free surfaces $z = \pm d/2$ (see Fig. 9.1). The loading on the curved boundary must act in the xy plane and be uniformly distributed in the z direction. According to (1.26) the surface tractions are, therefore, specified by

$$\bar{t}_x = \sigma_{xx} l + \tau_{xy} m, \qquad \bar{t}_y = \tau_{yx} l + \sigma_{yy} m, \qquad \bar{t}_z = 0 \qquad (9.1)$$

The body force component b_z must be zero, whilst b_x and b_y can be functions only of x and y.

For such problems, we can identify two distinct cases in which either the state of stress or the state of strain is two-dimensional.

9.2 PLANE STRESS

If the dimension d of the solid in the z direction is *very small* compared with the other dimensions, it follows that the conditions specifying stress-free plane surfaces given by

$$\sigma_{zz} = \tau_{zx} = \tau_{zy} = 0 \qquad (9.2)$$

may be assumed to hold throughout the thickness. This is an example of the two-dimensional stress state, *plane stress*, defined by the three components σ_{xx}, σ_{yy}

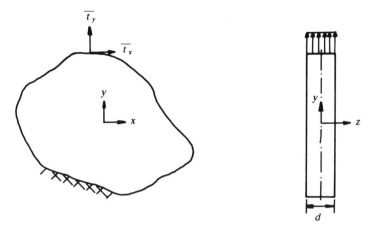

Fig. 9.1—Plane problem geometry.

and τ_{xy}. The corresponding strains in the xy plane are given by (3.2) as

$$e_{xx} = \frac{1}{E}(\sigma_{xx} - \nu\sigma_{yy}), \qquad e_{yy} = \frac{1}{E}(\sigma_{yy} - \nu\sigma_{xx}), \qquad \gamma_{xy} = \frac{\tau_{xy}}{G} \qquad (9.3)$$

Since there is also a direct strain in the z direction, given by $e_{zz} = -\nu(\sigma_{xx} + \sigma_{yy})$, it follows that the state of strain is three dimensional.

The body forces can usually be expressed in terms of a *potential function* Ω by

$$b_x = -\frac{\partial\Omega}{\partial x} \qquad \text{and} \qquad b_y = -\frac{\partial\Omega}{\partial y} \qquad (9.4)$$

in which case the equations of equilibrium (1.25) can be written as

$$\frac{\partial}{\partial x}(\sigma_{xx} - \Omega) + \frac{\partial\tau_{xy}}{\partial y} = 0$$

$$\frac{\partial\tau_{yx}}{\partial x} + \frac{\partial}{\partial y}(\sigma_{yy} - \Omega) = 0 \qquad (9.5)$$

Of the six equations of compatibility (2.26), all but the first one

$$\frac{\partial^2 e_{xx}}{\partial y^2} + \frac{\partial^2 e_{yy}}{\partial x^2} = \frac{\partial^2 \gamma_{xy}}{\partial x\,\partial y} \qquad (9.6)$$

are assumed to be automatically satisfied. In general, this will not be the case and a plane stress solution is, therefore, only an *approximate solution* of the three-dimensional equations of elasticity.

To satisfy equilibrium (9.5), we define the stresses in terms of a stress function Φ by

$$\sigma_{xx} = \frac{\partial^2\Phi}{\partial y^2} + \Omega, \qquad \sigma_{yy} = \frac{\partial^2\Phi}{\partial x^2} + \Omega, \qquad \tau_{xy} = -\frac{\partial^2\Phi}{\partial x\,\partial y} \qquad (9.7)$$

We can now express compatibility (9.6) first in terms of stresses, using (9.3), and then in terms of Φ, using (9.7), to give finally

$$\nabla^4\Phi + (1 - v)\,\nabla^2\Omega = 0 \tag{9.8}$$

where

$$\nabla^2 = \frac{\partial^2}{\partial x^2} + \frac{\partial^2}{\partial y^2}$$

for (x, y) coordinates

$$= \frac{\partial^2}{\partial r^2} + \frac{1}{r}\frac{\partial}{\partial r} + \frac{1}{r^2}\frac{\partial^2}{\partial \theta^2}$$

For a polar coordinate system (r, θ), where the body forces are defined in terms of a potential function Ω by

$$b_r = -\frac{\partial\Omega}{\partial r}, \qquad b_\theta = -\frac{1}{r}\frac{\partial\Omega}{\partial \theta}$$

the corresponding stresses are defined by

$$\sigma_{rr} = \frac{1}{r}\frac{\partial\Phi}{\partial r} + \frac{1}{r^2}\frac{\partial^2\Phi}{\partial \theta^2} + \Omega, \qquad \sigma_{\theta\theta} = \frac{\partial^2\Phi}{\partial r^2} + \Omega, \qquad \tau_{r\theta} = -\frac{\partial}{\partial r}\left(\frac{1}{r}\frac{\partial\Phi}{\partial \theta}\right) \tag{9.9}$$

9.3 PLANE STRAIN

In the case of plane strain the dimension d of the solid is *very large* compared with the other dimensions (see Fig. 9.2). It follows that the deformation of every slice of

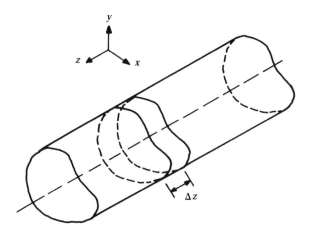

Fig. 9.2—Plane strain problem geometry.

thickness Δz away from the end surfaces can be assumed to be the same. The displacements u and v are, therefore, independent of z, whilst *plane cross-sections of the solid remain plane*. The latter condition implies that the axial displacement w is a function only of z. Since each slice must undergo the same change in thickness, we conclude that w must be a linear function of z: $w = \alpha z + \beta$, where α and β are constants.

For this simplified displacement field, it can be deduced from the strain–displacement equations (2.15) that $\gamma_{yz} = \gamma_{zx} = 0$, that e_{zz} is equal to the constant α and that the remaining strain components e_{xx}, e_{yy} and γ_{xy} are independent of z.

The axial stress on the end faces of the solid is given by the third of equations (3.2) as

$$\sigma_{zz} = Ee_{zz} + v(\sigma_{xx} + \sigma_{yy}) \tag{9.10}$$

Only when $v = 0$, or $\sigma_{xx} + \sigma_{yy}$ is independent of x and y, is it possible to determine a constant value e_{zz} for which σ_{zz} vanishes everywhere. However, we can always choose a value for e_{zz} such that there is no *resultant force* in the z direction on any cross-section of area A. This condition requires that

$$\int_A \sigma_{zz} \, dx \, dy = 0$$

and, after substituting for σ_{zz} from (9.10), we deduce that

$$e_{zz} = -\frac{\bar{\sigma}_{zz}}{E}$$

where

$$\bar{\sigma}_{zz} = \frac{v}{A} \int_A (\sigma_{xx} + \sigma_{yy}) \, dx \, dy$$

and hence

$$\sigma_{zz} = v(\sigma_{xx} + \sigma_{yy}) - \bar{\sigma}_{zz} \tag{9.11}$$

According to St Venant's principle (see section 4.4), any discrepancies between the free-end condition $\sigma_{zz} = 0$ and the actual value of σ_{zz} given by (9.11) will be confined to the neighbourhood of the end faces.

By superimposing an additional uniform stress $\bar{\sigma}_{zz}$, we can arrive at the plane strain condition $e_{zz} = 0$ (see Fig. 9.3). If, in a particular problem, the end constraints necessary to maintain this condition are not present, we shall need to superimpose the effect of a uniform stress $-\bar{\sigma}_{zz}$ on the plane strain solution.

With the aid of the plane strain condition, we can now eliminate σ_{zz} from the stress–strain relationships (3.2) to give

$$e_{xx} = \frac{1}{E^*}(\sigma_{xx} - v^*\sigma_{yy}), \qquad e_{yy} = \frac{1}{E^*}(\sigma_{yy} - v^*\sigma_{xx}), \qquad \gamma_{xy} = \frac{\tau_{xy}}{G} \tag{9.12}$$

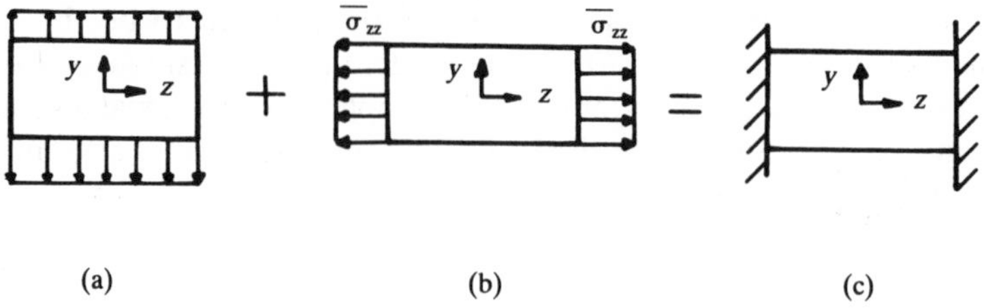

(a)   (b)   (c)

Fig. 9.3—(a) Constant-axial-strain problem $e_{zz} = -\bar{\sigma}_{zz}/E$, with (b) a uniform axial stress $\bar{\sigma}_{zz}$
superimposed to give (c) the plane strain condition $e_{zz} = 0$.

where

$$E^* = \frac{E}{1 - v^2} \quad \text{and} \quad v^* = \frac{v}{1 - v} \qquad (9.13)$$

These equations are seen to be identical with (9.3) for plane stress except that the
elastic constants E and v are now replaced by E^* and v^*.

Equilibrium reduces to the same pair of equations (9.5) as in the case of plane
stress, and compatibility is fully satisfied by the single equation (9.6), which can be
expressed in terms of the stress function as

$$\nabla^4 \Phi + (1 - v^*) \nabla^2 \Omega = 0 \qquad (9.14)$$

This is identical with (9.8) for plane stress except that v is now replaced by v^*. Plane
stress and plane strain are otherwise mathematically identical problems for the
same boundary conditions.

9.4 FINITE ELEMENT FORMULATION

In plane problems the displacement field is defined by the two components
$u = u(x, y)$ and $v = v(x, y)$, and the only undetermined stresses and strains are the
components in the xy plane. There is no unique way of modelling the geometry of
this plane using finite elements, and in practice a variety of element types can be
employed for this purpose. We shall confine our attention here to the simplest of
these, the straight-sided triangular element with three nodal points i, j and k, one at
each vertex (see Fig. 9.4). With two displacement degrees of freedom u and v for
each node the element has a total of *six* degrees of freedom in all.

9.4.1 Choice of element displacement functions
The criteria which determine the choice of displacement functions $[u]$ to describe
the displacement field within a particular element were outlined in section 6.7. The
first of these criteria requires the functions to be continuous within individual
elements and between adjacent elements. Use *polynomial functions* $[u]$ ensures

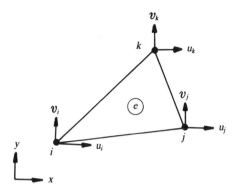

Fig. 9.4—Three-node triangular element for plane problems.

continuity within each element and, provided that u and v have a *linear variation* along each of the sides of the element, then continuity between adjacent elements is ensured. We are therefore lead to choose the following functions:

$$u = C_1 + C_2x + C_3y$$

$$v = D_1 + D_2x + D_3y$$

(9.15)

where the coefficients $C_1, D_1, \ldots$ are constants. The corresponding element strains are given by (2.14) as

$$[e] = \begin{bmatrix} e_{xx} \\ e_{yy} \\ \gamma_{xy} \end{bmatrix} = \begin{bmatrix} \dfrac{\partial u}{\partial x} \\ \dfrac{\partial v}{\partial y} \\ \dfrac{\partial u}{\partial y} + \dfrac{\partial v}{\partial x} \end{bmatrix} = \begin{bmatrix} C_2 \\ D_3 \\ C_3 + D_2 \end{bmatrix}$$

(9.16)

and all have constant values. For this reason the element is referred to as a *constant-strain triangle* (CST).

According to equations (2.17), rigid-body displacement in the xy plane is defined by

$$u = C_1 + C_3y, \qquad v = D_1 - C_3x$$

This is correctly modelled by $[u]$ when $C_2 = D_3 = 0$ and $D_2 = -C_3$, and the corresponding element strains given by (9.16) are all zero, as required by the third criterion. Although rigid-body displacement of individual elements is permissible, the complete assemblage of elements must be constrained by appropriate displacement boundary conditions to prevent rigid-body displacement of the mesh as a whole. Failure to do this will result in a singular system stiffness $[K]$ reflecting a lack of uniqueness in the solution of the equilibrium equations.

9.4.2 Element stiffness matrix

The polynomial coefficients in the displacement functions (9.15) are evaluated in terms of the nodal point displacements $[\delta^e] = [u_i \quad v_i \quad u_j \quad v_j \quad u_k \quad v_k]^T$ by requiring that

$$\begin{bmatrix} u_i \\ u_j \\ u_k \end{bmatrix} = \begin{bmatrix} 1 & x_i & y_i \\ 1 & x_j & y_j \\ 1 & x_k & y_k \end{bmatrix} \begin{bmatrix} C_1 \\ C_2 \\ C_3 \end{bmatrix} \quad \text{and} \quad \begin{bmatrix} v_i \\ v_j \\ v_k \end{bmatrix} = \begin{bmatrix} 1 & x_i & y_i \\ 1 & x_j & y_j \\ 1 & x_k & y_k \end{bmatrix} \begin{bmatrix} D_1 \\ D_2 \\ D_3 \end{bmatrix}$$

When the solution for the coefficients is substituted into (9.15), we obtain

$$\begin{bmatrix} u \\ v \end{bmatrix} = \begin{bmatrix} N_i & 0 & N_j & 0 & N_k & 0 \\ 0 & N_i & 0 & N_j & 0 & N_k \end{bmatrix} [\delta^e] \tag{9.17}$$

or

$$[u] = [N][\delta^e]$$

where the *shape functions* are defined by

$$N_i = \frac{1}{2A} (a_i + b_i x + c_i y), \text{ etc.} \tag{9.18}$$

in which A denotes the area of the element given by

$$A = \tfrac{1}{2} \det \begin{vmatrix} 1 & x_i & y_i \\ 1 & x_j & y_j \\ 1 & x_k & y_k \end{vmatrix} = \tfrac{1}{2}(a_i + a_j + a_k) \tag{9.19}$$

and

$$a_i = x_j y_k - x_k y_j, \qquad b_i = y_j - y_k, \qquad c_i = x_k - x_j \tag{9.20}$$

The remaining coefficients $a_j, b_j, \ldots$ are obtained from (9.20) by a *cyclic permutation* of the subscripts.

The element strains are found by substituting (9.17) in (9.16) to give

$$[e] = \begin{bmatrix} \dfrac{\partial N_i}{\partial x} & 0 & \dfrac{\partial N_j}{\partial x} & 0 & \dfrac{\partial N_k}{\partial x} & 0 \\[2mm] 0 & \dfrac{\partial N_i}{\partial y} & 0 & \dfrac{\partial N_j}{\partial y} & 0 & \dfrac{\partial N_k}{\partial y} \\[2mm] \dfrac{\partial N_i}{\partial y} & \dfrac{\partial N_i}{\partial x} & \dfrac{\partial N_j}{\partial y} & \dfrac{\partial N_j}{\partial x} & \dfrac{\partial N_k}{\partial y} & \dfrac{\partial N_k}{\partial x} \end{bmatrix} [\delta^e]$$

$$= \frac{1}{2A} \begin{bmatrix} b_i & 0 & b_j & 0 & b_k & 0 \\ 0 & c_i & 0 & c_j & 0 & c_k \\ c_i & b_i & c_j & b_j & c_k & b_k \end{bmatrix} [\delta^e] \tag{9.21}$$

or

$$[e] = [B][\delta^e]$$

For plane stress the stresses can be expressed in terms of the strains by solving equations (9.3) to give

$$
\begin{bmatrix} \sigma_{xx} \\ \sigma_{yy} \\ \tau_{xy} \end{bmatrix} = \frac{E}{1 - v^2} \begin{bmatrix} 1 & v & 0 \\ v & 1 & 0 \\ 0 & 0 & \frac{1}{2}(1 - v) \end{bmatrix} \begin{bmatrix} e_{xx} \\ e_{xy} \\ \gamma_{xy} \end{bmatrix}
\tag{9.22}
$$

or

$$[\sigma] = [D][e]$$

Since the strains $[e]$ are constant within each element, so too are the stresses $[\sigma]$.

The element stiffness matrix is given by (6.10) as

$$[K^e] = \int_{V_e} [B]^{\mathrm{T}}[D][B] \, \mathrm{d}V$$

and, since the matrices $[B]$ and $[D]$ are both constant over the volume V_e of the element, it follows that

$$[K^e] = A[B]^{\mathrm{T}}[D][B] \tag{9.23}$$

for a solid of unit thickness.

9.4.3 Stress boundary conditions

The equivalent set of nodal point forces for an element whose surface S_e carries prescribed tractions $[t]$ are, in general, given by (6.11) as

$$[F^e] = \int_{S_e} [N]^{\mathrm{T}}[t] \, \mathrm{d}S \tag{9.24}$$

In plane problems, S_e consists of only one edge mn of the element (see Fig. 9.5) and (9.24) becomes

$$[F^e_{mn}] = \int_0^l [N_{mn}]^{\mathrm{T}}[t_{mn}] \, \mathrm{d}\eta \tag{9.25}$$

where the coordinate η is measured along the edge mn of length l, and

$$[F^e_{mn}] = [P^e_m \quad Q^e_m \quad P^e_n \quad Q^e_n]^{\mathrm{T}}$$

$$[N_{mn}] = \begin{bmatrix} N_m & 0 & N_n & 0 \\ 0 & N_m & 0 & N_n \end{bmatrix}$$

$$N_m = 1 - \frac{\eta}{l} \quad \text{and} \quad N_n = \frac{\eta}{l}.$$

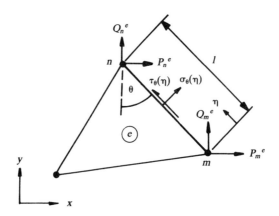

Fig. 9.5—Prescribed boundary stresses for a CST element.

Using (9.1), we can express the tractions in terms of the direct and shear stresses $\sigma_\theta = \sigma_\theta(\eta)$ and $\tau_\theta = \tau_\theta(\eta)$ on mn as

$$[t_{mn}] = \begin{bmatrix} t_x \\ t_y \end{bmatrix} = \begin{bmatrix} \sigma_\theta \cos\theta - \tau_\theta \sin\theta \\ \sigma_\theta \sin\theta + \tau_\theta \cos\theta \end{bmatrix} \tag{9.26}$$

where θ defines the inclination of the edge mn.

If the prescribed tractions vary *linearly* along mn, they can be expressed in terms of their values $[\bar{t}_{mn}]$ at nodes m and n by

$$[t_{mn}] = [N_{mn}][\bar{t}_{mn}] \tag{9.27}$$

where

$$[\bar{t}_{mn}] = [\bar{t}_{xm} \quad \bar{t}_{ym} \quad \bar{t}_{xn} \quad \bar{t}_{yn}]^\mathrm{T}$$

By substituting (9.27) into (9.25), we obtain

$$[F^e_{mn}] = \left(\int_0^l [N_{mn}]^\mathrm{T} [N_{mn}] \, \mathrm{d}\eta \right) [\bar{t}_{mn}] \tag{9.28}$$

which after the integration has been performed gives

$$\begin{bmatrix} P^e_m \\ Q^e_m \\ P^e_n \\ Q^e_n \end{bmatrix} = \frac{l}{6} \begin{bmatrix} 2 & 0 & 1 & 0 \\ 0 & 2 & 0 & 1 \\ 1 & 0 & 2 & 0 \\ 0 & 1 & 0 & 2 \end{bmatrix} \begin{bmatrix} \bar{t}_{xm} \\ \bar{t}_{ym} \\ \bar{t}_{xn} \\ \bar{t}_{yn} \end{bmatrix} \tag{9.29}$$

In the special case of *uniform prescribed tractions*, where $\bar{t}_{xm} = \bar{t}_{xn} = \bar{t}_x$ and $\bar{t}_{ym} = \bar{t}_{yn} = \bar{t}_y$, the nodal point forces are

$$P^e_m = P^e_n = \tfrac{1}{2}\bar{t}_x l \qquad \text{and} \qquad Q^e_m = Q^e_n = \tfrac{1}{2}\bar{t}_y l$$

and half the total load on the edge mn is assigned to each node.

9.4.4 Body forces

The equivalent set of nodal point forces for an element of volume V_e subjected to body forces $[b]$ per unit volume are given by (6.11) as

$$[F^e] = \int_{V_e} [N]^T[b] \, dV$$

For plane problems, with $[N]$ defined in (9.17) and $[b] = [b_x \quad b_y]^T$, it follows that

$$[F^e] = \int [b_x N_i \quad b_y N_i \quad b_x N_j \quad b_y N_j \quad b_x N_k \quad b_y N_k]^T \, dA \qquad (9.30)$$

If we move the origin of the (x, y) coordinate system to the centroid of the element, it can be shown that

$$\int N_i \, dA = \int N_j \, dA = \int N_k \, dA = \tfrac{1}{3}A \qquad (9.31)$$

In the case of *uniform body forces*, (9.30) can be integrated using (9.31) to give

$$[F^e] = \tfrac{1}{3}A[b_x \quad b_y \quad b_x \quad b_y \quad b_x \quad b_y]^T \qquad (9.32)$$

In other words, the resultant body force acting on the element is divided equally between the three nodes.

9.4.5 Element stresses

After the solution for the nodal point displacements $[\delta]$ has been obtained, there remains the task of calculating the corresponding stresses in each element. Using equations (9.21) and (9.22), we can deduce that these stresses are given in terms of the nodal point displacements $[\delta^e]$ for the element by

$$[\sigma] = [D][B][\delta^e] \qquad (9.33)$$

and are *uniform* over the element.

9.4.6 Use of FIESTA2

To select the appropriate element subroutine ELEM3 in FIESTA2, we put IETYPE=3 in the data file specifying the finite element model (see Fig. 7.2). The automatic mesh generation facility (see section 7.3) is available if required.

Subroutine ELEM3 calculates the element stiffness array SE using (9.23), and the array FE of nodal point forces corresponding to the combined effect of linearly varying boundary stresses σ_θ, τ_θ and uniform body forces b_x, b_y using (9.29) and (9.32), respectively. The array of element stresses SS is calculated in subroutine STR3 using (9.33).

The material data for the problem are specified in the manner described in section 7.2.2. The NOPROP=4 properties for each property set ISET in a plane stress problem are stored in the array EDAT as follows:

$$EDAT(ISET,1) = E, \qquad EDAT(ISET,2) = v$$

$$EDAT(ISET,3) = b_x, \qquad EDAT(ISET,4) = b_y$$

In the case of plane strain the elastic constants E and v must be replaced by E^* and v^* defined in (9.13).

Stress boundary conditions are specified as described in section 7.2.3. When the displacement boundary conditions are specified, it is essential to ensure that they are sufficient in number to prevent $[K]$ from being a singular matrix. Failure to do so will result in the error message 'ILL-CONDITIONED EQUATIONS' when FIESTA2 is run.

9.5 DEEP BEAM ANALYSIS

In the elementary theory of the bending of beams, it is assumed that plane cross-sections of a beam remain plane during bending. No account is taken of the presence of shear stresses in the beam which cause plane cross-sections to *warp*. The effect of this warping on the stress and displacement fields becomes significant in the case of a *deep beam* where the depth of the cross-section is a substantial proportion of the length of the beam.

The simply supported deep beam shown in Fig. 9.6 has a narrow rectangular cross-section, of breadth b and depth $2c$, and carries a uniformly distributed load of intensity q per unit length. We obtain first an analytical solution for the plane stress problem using the stress function approach and later compare this with a finite element approximation.

9.5.1 Analytical solution

The solution for the stress function $\Phi(x, y)$ is assumed to be in the form of a polynomial in x and y. This solution must satisfy strain compatibility (9.8) with

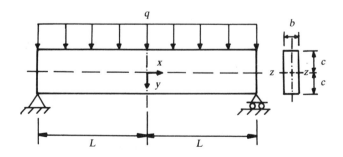

Fig. 9.6—Simply supported deep beam carrying a uniformly distributed load.

$\Omega = 0$ and give stresses defined by (9.7) which satisfy the boundary conditions

$$\tau_{xy} = 0 \qquad\qquad \text{at } y = \pm c$$

$$\sigma_{yy} = 0 \qquad\qquad \text{at } y = +c$$

$$\sigma_{yy} = -q/b \qquad\qquad \text{at } y = -c \qquad\qquad (9.34)$$

$$\int_{-c}^{+c} \tau_{xy} b \, dy = \pm\tfrac{1}{2}qL \qquad \text{at } x = \pm L$$

$$\int_{-c}^{+c} \sigma_{xx} by \, dy = 0 \qquad \text{at } x = \pm L$$

Assuming that the solution for Φ is a complete *fifth-order* polynomial whose terms are given by Pascal's triangle (see Fig. 6.2), we can eliminate most of the total of 21 terms in such a polynomial on the basis of the following observations.

(1) The constant term and the terms linear in x and y will give zero stresses and need not be included.
(2) The resultant axial force must be zero on any cross-section of the beam. It follows that σ_{xx} and hence Φ are *odd* functions of y.
(3) Since σ_{yy} is independent of x on the lower and upper surfaces, according to the boundary conditions (9.34), Φ must not contain powers of x greater than x^2.
(4) The cross-section $x = 0$ is a plane of symmetry and thus Φ is an *even* function of x.

We are therefore left with only the five terms

$$\Phi = C_1 x^2 + C_2 x^2 y + C_3 y^3 + C_4 x^2 y^3 + C_5 y^5$$

where $C_1, C_2, \ldots$ are constants. This solution satisfies compatibility (9.8) if $C_4 = -5C_5$, and the corresponding stresses are given by (9.7) as

$$\sigma_{xx} = \frac{\partial^2 \Phi}{\partial y^2} = 6C_3 y - 30C_5 x^2 y + 20C_5 y^3$$

$$\sigma_{yy} = \frac{\partial^2 \Phi}{\partial x^2} = 2C_1 + 2C_2 y - 10C_5 y^3$$

$$\tau_{xy} = -\frac{\partial^2 \Phi}{\partial x \, \partial y} = -2C_2 x + 30C_5 xy^2$$

The first three boundary conditions (9.34) require that

$$-2C_2 + 30C_5 c^2 = 0$$

$$2C_1 + 2C_2 c - 10C_5 c^3 = 0$$

$$2C_1 - 2C_2 c + 10C_5 c^3 = -q/b$$

from which we obtain $C_1 = -qc^3/6I$, $C_2 = qc^2/4I$ and $C_5 = q/60I$, where $I = 2bc^3/3$ is the second moment of area of the cross-section about the neutral axis zz. The remaining pair of boundary conditions are satisfied for $C_3 = (q/12I)(L^2 - 2c^2/5)$. Hence the solution for the stresses is given by

$$\sigma_{xx} = \frac{q}{2I}\left(L^2 - x^2\right)y + \frac{q}{2I}\left(\frac{2y^3}{3} - \frac{2c^2y}{5}\right)$$

$$\sigma_{yy} = \frac{-q}{2I}\left(\frac{y^3}{3} - c^2y + \frac{2c^3}{3}\right) \tag{9.35}$$

$$\tau_{xy} = \frac{-q}{2I}(c^2 - y^2)x$$

where the first term in the solution for σ_{xx} represents the solution obtained using elementary beam theory. The second term results from the warping of cross-sections and is negligible for a beam whose span is large in comparison with its depth.

The solutions for the stresses are exact only when they give distributions for the normal and shear stresses on the end faces $x = \pm L$ which coincide with those actually prescribed. According to St Venant's principle (see section 4.4), however, any discrepancies here will have a negligible effect on the stresses in the region away from the end faces. This follows from the fact that there are no discrepancies in the stress resultants on these faces.

The displacements are found by substituting the stresses from (9.35) into the strain–displacement equations (2.14) and integrating. The resulting solution for the lateral deflection at the centre of the beam $x = y = 0$ is given by

$$\delta = \frac{5}{24}\frac{qL^4}{EI}\left[1 + \frac{12}{5}\frac{c^2}{L^2}\left(\frac{4}{5} + \frac{v}{2}\right)\right] \tag{9.36}$$

where the first term represents the deflection given by the elementary theory of bending. The second term accounts for the effect of the shear stresses and is of significant magnitude for a deep beam.

9.5.2 Finite element solution
Consider now a finite element solution to the problem for $b = L/15, c = L/3$ and $v = 0.3$. Because the midspan cross-section of the beam is a plane of symmetry for the problem, it is only necessary to model one-half of the beam. To illustrate the convergence of the finite element method, the three uniform meshes based on grids of 3 by 4, 5 by 7, and 9 by 13 nodal points (see Fig. 9.7) are used. The nodes are numbered along the depth of the beam to minimise the semi-bandwidth of $[K]$.

The accuracy of the finite element results can be assessed by comparing the vertical deflection δ at the centre of the beam with the exact solution given by (9.36). In Fig. 9.8, it is seen that the finite element results for the dimensionless deflection $\delta E/q$ (see section 4.7) converge towards the exact solution as the number of degrees of freedom in the model is increased.

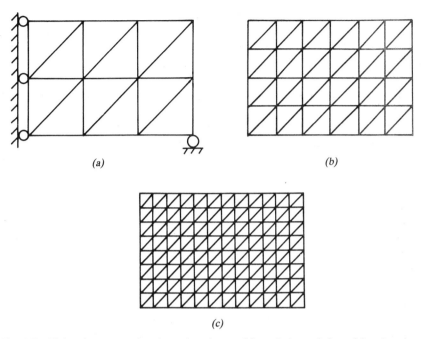

Fig. 9.7—Finite element meshes for a deep beam: (a) mesh 1 consisting of 3 × 4 nodes:
(b) mesh 2 consisting of 5 × 7 nodes: (c) mesh 3 consisting of 9 × 13 nodes.

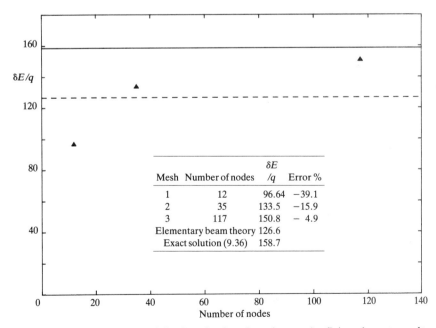

Mesh	Number of nodes	$\delta E/q$	Error %
1	12	96.64	−39.1
2	35	133.5	−15.9
3	117	150.8	− 4.9
Elementary beam theory		126.6	
Exact solution (9.36)		158.7	

Fig. 9.8—Maximum lateral deflection δ of a deep beam: ▲, finite element results;
——, exact solution; – – –, elementary beam theory.

In section 5.4, we saw that the discretisation error associated with the use of bar elements is of the order of l^2, where l is the element length. It can be shown that, in general, the use of conforming elements results in errors of the order of l^{N+1}, where l is the side length of a typical element and N is the degree of the polynomials used in the displacement functions $[u]$. Since the stresses are first derivatives of the displacements, it follows that the corresponding discretisation error in their approximation is of the order of l^N.

For CST elements, $N = 1$ and the error for the displacements is of the order of l^2, as in the case of the bar elements. It follows that the extrapolation formula (5.27) can be used to obtain improved approximations, for example $\delta E/q = 145.73$ using the results for meshes 1 and 2, and $\delta E/q = 156.58$ using the results for meshes 2 and 3. The latter estimate is within 2% of the exact solution.

The stresses predicted by the finite element method are uniform within each element and, therefore, undergo *step discontinuities* at the inter-element boundaries. It is usual to associate the stress in a particular element with the centroid of that element. Commercial finite element programs employ various systematic *smoothing procedures* to aid the interpretation of the solution for the stresses. These include calculating the stresses at each nodal point by averaging their values over those elements sharing that particular node.

The distribution of the dimensionless bending stress $\sigma_{xx}L/q$ for elements adjacent to the midspan cross-section $x = 0$, obtained using mesh 3, is compared with the exact solution (9.35) in Fig. 9.9.

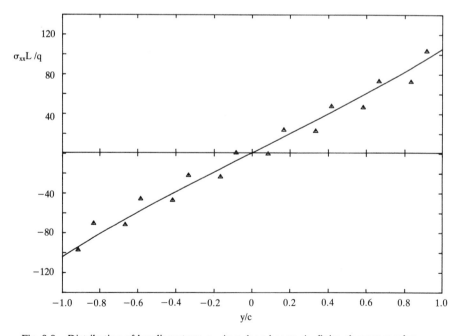

Fig. 9.9—Distribution of bending stress σ_{xx} in a deep beam: ▲, finite element results; —, exact solution (9.35).

9.6 EFFECT OF CONCENTRATED LOADS

When a concentrated load is applied at a point on the surface of a solid, the resulting stress distribution will have large gradients in the vicinity of this point. The effect of these *stress concentrations* may produce stresses which are sufficiently high locally to cause the material to yield, whilst the bulk of the solid is behaving elastically.

9.6.1 Analytical solutions

Consider the effect of a concentrated vertical line load of intensity P per unit length acting on the horizontal straight boundary of a semi-infinite elastic plate (see Fig. 9.10). Using polar coordinates, we assume a solution for the stress function of the *separable* form $\Phi = rF(\theta)$ which satisfies compatibility (9.8) provided that

$$\frac{\mathrm{d}^4F}{\mathrm{d}\theta^4} + 2\,\frac{\mathrm{d}^2F}{\mathrm{d}\theta^2} + F = 0$$

We require the *symmetric* solution

$$F(\theta) = A\cos\theta + B\theta\sin\theta$$

for which the corresponding stresses given by (9.9) are

$$\sigma_{rr} = \frac{2B}{r}\cos\theta, \qquad \sigma_{\theta\theta} = \tau_{r\theta} = 0 \tag{9.37}$$

These satisfy the boundary conditions $\sigma_{\theta\theta} = \tau_{r\theta} = 0$ at $\theta = \pm\pi/2$ and, to evaluate the constant B, we use the equation of equilibrium for the vertical forces acting on a semi-circular section of the plate of radius r (see Fig. 9.10) given by

$$P = -\sigma_{rr}\int_{-\pi/2}^{+\pi/2}\cos\theta\, r\,\mathrm{d}\theta$$

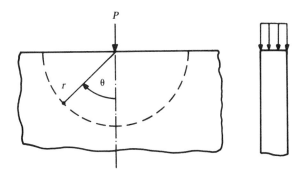

Fig. 9.10—Semi-infinite plate carrying a line load.

After integration, we obtain $B = -P/\pi$ and, therefore, the only nonzero stress in (9.37) can be written as

$$\sigma_{rr} = -\frac{2P}{\pi r} \cos \theta \qquad\qquad (9.38)$$

This is a *radial stress field* with a singularity at the point of load application $r = 0$ where the stress σ_{rr} is theoretically infinite. The singularity results from our assumption that the load acts over an infinitesimally small area and that the material behaves elastically. In practice the load will be distributed over a small but finite area, and plastic deformation will occur locally to relieve the high stresses.

Next, we examine the problem of a circular elastic disc compressed by two loads P along a diameter AB (see Fig. 9.11). To begin with, we assume that the loads at A and B produce radial stress fields at any point E given by (9.38) as

$$\sigma_{r_1r_1} = -\frac{2P}{\pi r_1} \cos \theta_1 \qquad \text{and} \qquad \sigma_{r_2r_2} = -\frac{2P}{\pi r_2} \cos \theta_2$$

respectively. At a point E' on the circumference of the disc, r_1 and r_2 are orthogonal coordinates and

$$\frac{\cos \theta_1}{r_1} = \frac{\cos \theta_2}{r_2} = \frac{1}{d}$$

where d is the diameter of the disc. When the two loads act together, $\sigma_{r_1r_1}$ and $\sigma_{r_2r_2}$ are the two principal stresses at E' and are each equal to $-2P/\pi d$. The circumference of the disc therefore carries a uniform compressive stress $2P/\pi d$ which is removed by superimposing the additional stress field due to a tensile stress

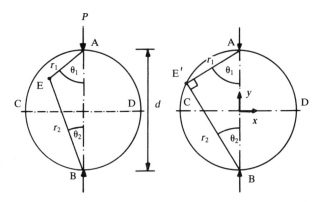

Fig. 9.11—Circular disc compressed by a pair of diametral loads.

of the same magnitude. On the diametral plane CD, we have the resulting direct stress

$$\sigma_{yy} = -\frac{4P}{\pi r_1} \cos^3 \theta_1 + \frac{2P}{\pi d}$$

$$= \frac{2P}{\pi d}\left[1 - \frac{4d^4}{(d^2 + 4x^2)^2}\right]$$

(9.39)

which has the maximum compressive value of $-6P/\pi d$ at the centre of the disc. At the ends of the diameter, σ_{yy} vanishes. Along the diameter AB, we have

$$\sigma_{yy} = \frac{-2P}{\pi r_1} - \frac{2P}{\pi r_2} + \frac{2P}{\hbar d}$$

$$= \frac{2P}{\pi d}\left(1 - \frac{4d^2}{d^2 - 4y^2}\right)$$

(9.40)

which has *singularities* at the points of load application $y = \pm d/2$.

9.6.2 Finite element solution

Since the disc has two axes of symmetry $x = 0$ and $y = 0$, we need only model one quadrant and impose symmetry conditions along the diametral boundaries of the

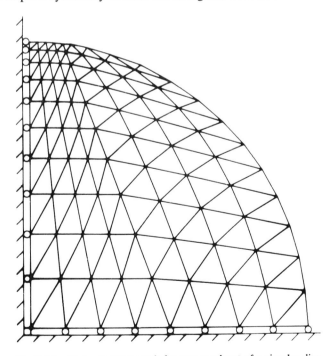

Fig. 9.12—Finite element mesh for one quadrant of a circular disc.

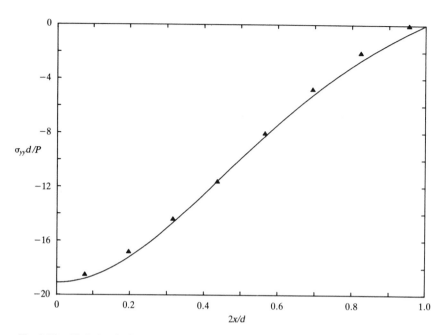

Fig. 9.13—Variation in the stress component σ_{yy} along the diameter $y = 0$ in a circular disc: ▲, finite element results; —, exact solution (9.39).

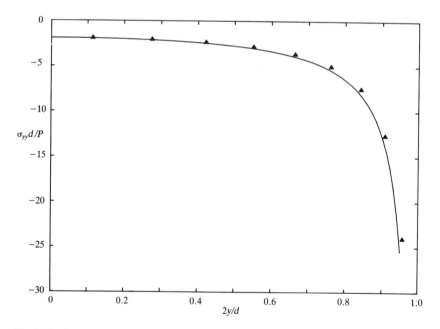

Fig. 9.14—Variation on the stress component σ_{yy} along the diameter $x = 0$ in a circular disc: ▲, finite element results; —, exact solution (9.40).

finite element mesh. The quadrant is divided up into two subregions each of which is modelled by defining coordinate mappings for a uniform square mesh based on an 11 by 5 grid of nodes (see Fig. 9.12). The element size is reduced in the region around the point of load application where the highest stress gradients are anticipated. This is achieved by adjusting the positions of the side points used to define the mapping functions (see section 7.3.3).

The finite element results for the variation in σ_{yy} along the two diameters $y = 0$ and $x = 0$ are shown in Figs 9.13 and 9.14, respectively. Over the entire length of the diameter $y = 0$ the results are close to the exact solution (9.39), whilst along $x = 0$ the results are close to the exact solution (9.40) for y/d less than 0.8. The stress field in the region adjacent to the singularity is, predictably, not well represented with CST elements.

9.7 STRESS INTENSITY FACTOR FOR A CRACK

The calculation of the stress intensity factor is an essential step in the fracture mechanics analysis of cracks in engineering components (see section 3.5). Analytical solutions are available for only a very limited range of crack problems, mostly involving infinite plate geometries with very simple loading. Before examining the application of the finite element method to this problem, we derive analytically the form of the stress and displacement fields adjacent to the crack tip.

9.7.1 Eigenfunction expansion method

We begin by assuming that the stress function for a crack whose faces are defined by $\theta = \pm\pi$ (see Fig. 3.7) is of the separable form $\Phi = r^{\lambda+1}F(\theta)$, where λ is an undetermined constant. To find the function F, we substitute this solution into the strain compatibility equation (9.8) to give

$$\frac{d^4F}{d\theta^4} + [(\lambda - 1)^2 + (\lambda + 1)^2]\frac{d^2F}{d\theta^2} + (\lambda - 1)^2(\lambda + 1)^2F = 0$$

and, for mode I loading, we take the *symmetric* solution

$$F(\theta) = A\,\cos[(\lambda - 1)\theta] + B\,\cos[(\lambda + 1)\theta] \tag{9.41}$$

where A and B are constants. Using (9.9), we can write the stresses in terms of F as

$$\sigma_{rr} = \frac{1}{r}\frac{\partial\Phi}{\partial r} + \frac{1}{r^2}\frac{\partial^2\Phi}{\partial\theta^2} = r^{\lambda-1}\left[\frac{d^2F}{d\theta^2} + (\lambda + 1)F\right]$$

$$\sigma_{\theta\theta} = \frac{\partial^2\Phi}{\partial r^2} = (\lambda + 1)\lambda r^{\lambda-1}F \tag{9.42}$$

$$\tau_{r\theta} = -\frac{\partial}{\partial r}\left(\frac{1}{r}\frac{\partial\Phi}{\partial\theta}\right) = -\lambda r^{\lambda-1}\frac{dF}{d\theta}$$

and the stress boundary conditions $\sigma_{\theta\theta} = \tau_{r\theta} = 0$ on the crack faces $\theta = \pm\pi$ as $F = dF/d\theta = 0$. Substituting for F from (9.41), we can express these two conditions as

$$A \cos[(\lambda - 1)\pi] + B \cos[(\lambda + 1)\pi] = 0$$

$$A(\lambda - 1) \sin[(\lambda - 1)\pi] + B(\lambda + 1) \sin[(\lambda + 1)\pi] = 0$$

which will have *non-trivial solutions* A and B provided that

$$\begin{vmatrix} \cos[(\lambda - 1)\pi] & \cos[(\lambda + 1)\pi] \\ (\lambda - 1) \sin[(\lambda - 1)\pi] & (\lambda + 1) \sin[(\lambda + 1)\pi] \end{vmatrix} = 0.$$

This condition leads to the *eigenequation* $\sin 2\lambda\pi = 0$ which has an infinite set of positive solutions $\lambda = \frac{1}{2}, 1, 1\frac{1}{2}, \ldots$. Associated with each *eigenvalue* λ_i there is an *eigenfunction* $r^{\lambda_i+1}F_i$, so that the general solution for Φ can be written as

$$\Phi = \sum_{i=1}^{\infty} r^{\lambda_i+1}\{A_1 5A_1 \cos[(\lambda_i - 1)\theta] + B_i \cos[(\lambda_i + 1)\theta]\} \tag{9.43}$$

Close to the crack tip we require only the first term in this *eigenfunction expansion* for which the corresponding stresses are

$$\sigma_{rr} = \frac{K_I}{\sqrt{2\pi r}} \left[\frac{5}{4} \cos\left(\frac{\theta}{2}\right) - \frac{1}{4} \cos\left(\frac{3\theta}{2}\right) \right]$$

$$\sigma_{\theta\theta} = \frac{K_I}{\sqrt{2\pi r}} \left[\frac{3}{4} \cos\left(\frac{\theta}{2}\right) + \frac{1}{4} \cos\left(\frac{3\theta}{2}\right) \right] \tag{9.44}$$

$$\tau_{r\theta} = \frac{K_I}{\sqrt{2\pi r}} \left[\frac{1}{4} \sin\left(\frac{\theta}{2}\right) + \frac{1}{4} \sin\left(\frac{3\theta}{2}\right) \right]$$

and the plane stress components of displacement in the x and y directions are

$$u = \frac{K_I}{2E}\sqrt{\frac{r}{2\pi}} \left[(5 - 3v) \cos\left(\frac{\theta}{2}\right) - (1 + v) \cos\left(\frac{3\theta}{2}\right) \right]$$

$$v = \frac{K_I}{2E}\sqrt{\frac{r}{2\pi}} \left[(7 - v) \sin\left(\frac{\theta}{2}\right) - (1 + v) \sin\left(\frac{3\theta}{2}\right) \right] \tag{9.45}$$

These solutions are identical for all mode I crack problems to within the single parameter $K_I = A_1\sqrt{2\pi}$, the stress intensity factor. The value of K_I may be approximated numerically by the *boundary collocation technique* in which a finite number of terms in the expansion (9.43) for Φ are used to satisfy the remaining boundary conditions on the edges of the plate remote from the crack faces.

9.7.2 Finite element solution
We now employ the finite element method to estimate the stress intensity factor for a central crack of length $2a$ in a rectangular plate of finite width $2b$ which is loaded

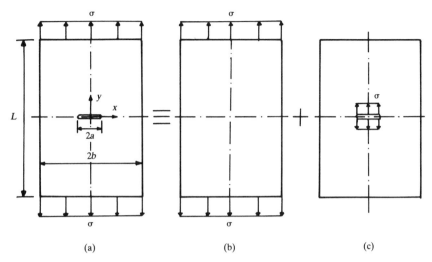

Fig. 9.15—(a) Crack problem for a finite width plate loaded by a uniform tensile stress σ, represented as the summation of the problems for (b) case 1 (the loaded plate without the crack) and (c) case 2 (the plate with pressure loading σ on the crack faces).

by a uniform tensile stress σ (see Fig. 9.15). According to the principle of superposition (see section 4.5), the solution for the stresses in the plate is found by summing the stress field in the plate without the crack (case 1), with the stress field in the plate when the crack faces carry a uniform pressure loading $\sigma_{yy} = -\sigma$ (case 2). The solution for K_I in the original problem is equal to K_I in case 2 since the stress intensity factor is zero for case 1.

The problem is completely symmetric about both the x and the y axes and hence we need only model one quadrant of the plate. Small elements are required in the vicinity of the crack tip where the stress gradients are expected to be large, whilst relatively large elements are adequate for the region remote from the crack. This is achieved using FIESTA2 by generating a graded square mesh of elements (see section 7.3.2) and then fitting this mesh to the rectangular region occupied by the quadrant using suitable coordinate mappings. The resulting mesh, shown in Fig. 9.16, has 89 nodes and 140 elements. To increment the crack length, we simply release the nodal point at the current crack tip by amending the boundary conditions accordingly.

There are two commonly used methods for calculating the stress intensity factor from the results of a finite element analysis. Firstly, K_I may be found directly by correlating the predicted stresses or displacements adjacent to the crack tip with the singular analytical solutions. The crack face displacements $v(r, \pi)$ are normally used for this purpose, and K_I is deduced by expressing the second of equations (9.45) as

$$K_I = \tfrac{1}{4}\sqrt{2\pi} \ \ E \lim_{r \to 0} \left[\frac{v(r, \pi)}{\sqrt{r}} \right] \tag{9.46}$$

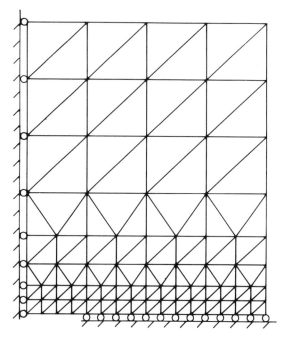

Fig. 9.16—Finite element mesh for one quadrant of a cracked plate.

If the exact solution for v is used with (9.46) to plot the estimated value of K_I as a function of r, then the exact value is given by the intercept of the curve at $r = 0$. For example, in the case of a crack of length $2a$ in an infinite plate carrying a tensile stress σ (see Fig. 3.7) the exact solution for the crack face displacement is

$$v = \frac{2\sigma}{E} \sqrt{(2ar - r^2)}$$

so that (9.46) becomes

$$K_I = \sigma\sqrt{\pi a} \lim_{r \to 0} \left[1 - \frac{1}{4}\left(\frac{r}{a}\right) - \frac{1}{32}\left(\frac{r}{a}\right)^2 - \cdots \right]$$

and the curve has an almost constant slope for $r/a < 1$ (see Fig. 9.17). The finite element results for v close to the crack tip are inaccurate owing to the inability of the elements to represent the stress singularity. However, the curve approaches a constant slope as r increases, and a good estimate of K_I is obtained by extrapolating this constant slope back to $r = 0$. In Fig. 9.18 the results obtained by this method for a range of crack lengths are compared with the boundary collocation solution which is very closely approximated by the formula

$$K_I = \sigma\sqrt{\pi a} \sqrt{\sec\left(\frac{\pi a}{2b}\right)} \tag{9.47}$$

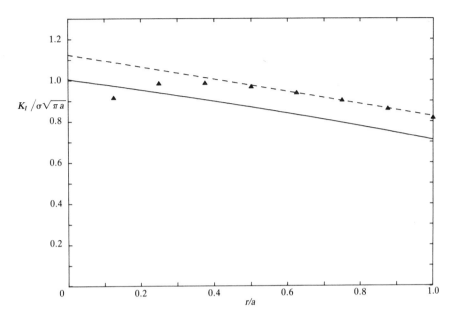

Fig. 9.17—Extrapolation method for estimating K_I: ▲, finite element results for $a/b = 0.5$; ---, extrapolation line; ——, exact solution (9.46) for an infinite plate.

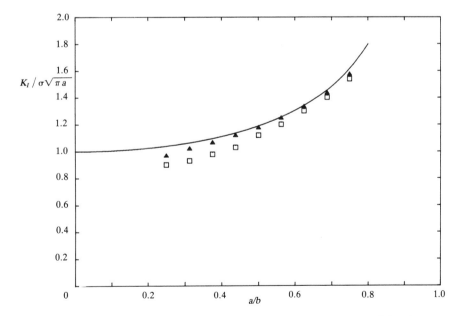

Fig. 9.18—Variation is stress intensity factor K_I with the crack length: ▲, finite element results, energy method; □, finite element results, extrapolation method; ——, exact solution (9.47).

The greatest error occurs for short cracks where the number of nodal displacements on the crack face is too small to allow reliable extrapolation.

In the second method, K_I is obtained indirectly by way of an energy calculation for $\mathcal{G}_I$. With the area of the crack face given by $A = 2a$ for a plate of unit thickness, we can write (3.22) as

$$\mathcal{G}_I = \frac{1}{2}\left(\frac{\partial W}{\partial A}\right)_\sigma = \frac{1}{4}\left(\frac{\partial W}{\partial a}\right)_\sigma = \left(\frac{\partial W'}{\partial a}\right)_\sigma \qquad (9.48)$$

$W' = W/4$ is the work done by the pressure forces $\sigma_{yy} = -\sigma$ acting on the crack face for one quadrant of the plate where

$$W' = -\int_0^a \sigma_{yy} v \ \mathrm{d}x = \sigma \int_0^a v \ \mathrm{d}x \qquad (9.49)$$

For CST elements, **v** is a linear function of x and the exact value of the integral in (9.49) is given by the trapezoidal rule as

$$W' = \sigma(\Sigma\, v - \tfrac{1}{2}v_0)\, \Delta x \qquad (9.50)$$

where v_0 is the crack face displacement at $x = 0$, and the summation Σ is performed over all the crack face nodes. Δx is the length of the element sides forming the crack face. The value of $\mathcal{G}_I$ given by (9.48) is found by differentiating numerically the finite element results for W' shown in Fig. 9.19. Finally, K_I is given by (3.30) as

$$K_I = \sqrt{E\,\mathcal{G}_I}$$

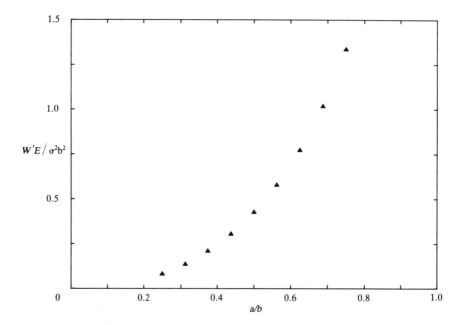

Fig. 9.19—Work done W' on one quadrant of the cracked plate by the crack face pressure σ.

The finite element estimates of K_1 shown in Fig. 9.18 are significantly better than those obtained using the extrapolation method. Again the results are least accurate for short cracks whilst the error is as low as 1% for long cracks.

9.8 STRESS CONCENTRATION DUE TO A HOLE

When a hole is drilled in a plate carrying a tensile stress σ (see Fig. 9.20), there is a localised disturbance of the uniform stress field. Unlike the problems discussed in sections 9.6 and 9.7 the stresses in this problem remain finite. However, near the surface of the hole the *stress concentration factor*, defined as the ratio of the maximum tensile stress to the tensile stress remote from the hole, may still be sufficiently large to cause plastic deformation. We examine first the analytical solution for an infinite plate and later the finite element solution for a finite plate.

9.8.1 Analytical solution for infinite plate problem

The uniform stress state in an infinite plate without a hole is defined by $\sigma_{xx} = \tau_{xy} = 0$, $\sigma_{yy} = \sigma$ and corresponds to the stress function $\Phi_1 = \frac{1}{2}\sigma x^2$. For a polar coordinate system, Φ_1 can be written as

$$\Phi_1 = \frac{1}{2}\sigma r^2 (1 + \cos 2\theta) \tag{9.51}$$

and the stress state becomes

$$(\sigma_{rr})_1 = \frac{1}{2}\sigma(1 - \cos 2\theta), \quad (\sigma_{\theta\theta})_1 = \frac{1}{2}\sigma(1 + \cos 2\theta), \quad (\tau_{r\theta})_1 = \frac{1}{2}\sigma \sin 2\theta \tag{9.52}$$

When a hole of radius a is made in the plate, the boundary conditions on its surface are $\sigma_{rr} = \tau_{r\theta} = 0$, for all values of θ. To satisfy these conditions, we must add to the stress function Φ_1 defined in (9.51) a second stress function Φ_2 for which the values of σ_{rr} and $\tau_{r\theta}$ at $r = a$ are equal and opposite to those given by (9.52). In addition, the stresses corresponding to Φ_2 must be negligible for $r \gg a$. The stress

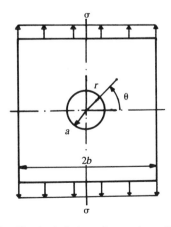

Fig. 9.20—Circular hole in a plate under uniform tension.

function with these properties is given by

$$\Phi_2 = -\tfrac{1}{2}\sigma a^2 \left[\log r + \left(1 - \frac{a^2}{2r^2} \right) \cos 2\theta \right]$$

and the corresponding stresses are

$$(\sigma_{rr})_2 = -\frac{\sigma a^2}{2r^2}\left[1 - \left(4 - \frac{3a^2}{r^2} \right) \cos 2\theta \right]$$

$$(\sigma_{\theta\theta})_2 = \frac{\sigma a^2}{2r^2}\left(1 + \frac{3a^2}{r^2} \cos 2\theta \right)$$

$$(\tau_{r\theta})_2 = \frac{\sigma a^2}{2r^2}\left(2 - \frac{3a^2}{r^2} \right) \sin 2\theta$$

The solution that we seek is obtained by adding the stress fields corresponding to Φ_1 and Φ_2 to give

$$\sigma_{rr} = (\sigma_{rr})_1 + (\sigma_{rr})_2$$
$$= \tfrac{1}{2}\sigma\left(1 - \frac{a^2}{r^2} \right)\left[1 - \left(1 - \frac{3a^2}{r^2} \right) \cos 2\theta \right]$$

$$\sigma_{\theta\theta} = (\sigma_{\theta\theta})_1 + (\sigma_{\theta\theta})_2$$
$$= \tfrac{1}{2}\sigma\left[1 + \frac{a^2}{r^2} + \left(1 + \frac{3a^4}{r^4} \right) \cos 2\theta \right] \tag{9.53}$$

$$\tau_{r\theta} = (\tau_{r\theta})_1 + (\tau_{r\theta})_2$$
$$= \tfrac{1}{2}\sigma\left(1 - \frac{a^2}{r^2} \right)\left(1 + \frac{3a^2}{r^2} \right) \sin 2\theta$$

Along the plane of symmetry $\theta = 0$, we have the tensile stress

$$\sigma_{\theta\theta} = \tfrac{1}{2}\sigma\left(2 + \frac{a^2}{r^2} + \frac{3a^4}{r^4} \right) \tag{9.54}$$

which has a maximum value of 3σ at the surface of the hole $r = a$ and rapidly approaches the uniform value σ as r increases.

9.8.2 Finite element solution for finite plate problem

One quadrant of a square plate, of side length $2b$, with a hole of radius $a = 0.1b$ is divided into two subregions each of which is modelled using a uniform square mesh, based on an 11 by 5 grid of nodes, and appropriate mapping functions. The mesh shown in Fig. 9.21 has 99 nodes and 160 elements whose size is reduced in the region adjacent to the surface of the hole where the largest stress gradients are expected. Use of small elements around the edge of the hole also facilitates the accurate modelling of the circular shape of this boundary.

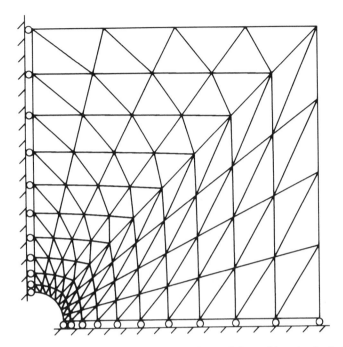

Fig. 9.21—Finite element mesh for one quadrant of plate with a circular hole.

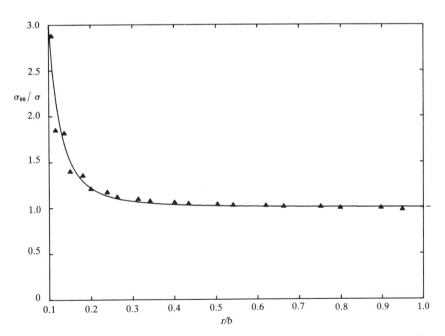

Fig. 9.22—Variation in the stress component $\sigma_{\theta\theta}$ alone $\theta = 0$ for a plate with a circular hole:
▲, finite element results; ——, exact solution (9.54).

The results shown in Fig. 9.22 for the tensile stress $\sigma_{\theta\theta}$ along $\theta = 0$ are seen to be in close agreement with the exact infinite plate solution (9.54). An even better resolution of the stress concentration factor can be achieved by increasing the degree of refinement of the mesh.

As the width of the plate is reduced, so the exact solution for the stresses will deviate increasingly from the infinite plate solution. The larger the ratio a/b, the greater will be the effect of the finite boundaries on the stress concentration.

9.9 STRESS CONCENTRATION AT A FILLET

Sudden changes in the cross-section of a plate, involving sharp corners, will give rise to very high stress concentrations. The problem is lessened by the introduction of a *fillet radius R* (see Fig. 9.23) to blend one section smoothly into the next. The stress concentration at such a fillet radius is not amenable to analytical treatment.

A problem of this type can be analysed with the finite element method by modelling one-half of the plate using the mesh shown in Fig. 9.24. This mesh corresponds to the one used for the plate with a circular hole with an additional subregion of elements added.

The variation in the maximum shear stress τ_{max} along the boundary of the plate is shown in Fig. 9.25. The transition between uniform stress states $\tau_{max} = \frac{1}{2}\sigma$ at one

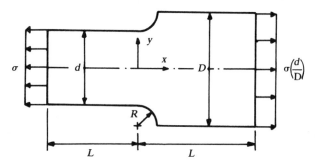

Fig. 9.23—Plate under tension with a sudden change in cross-section area.

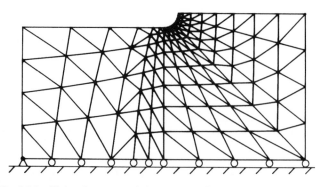

Fig. 9.24—Finite element mesh for one half of a plate with a fillet radius.

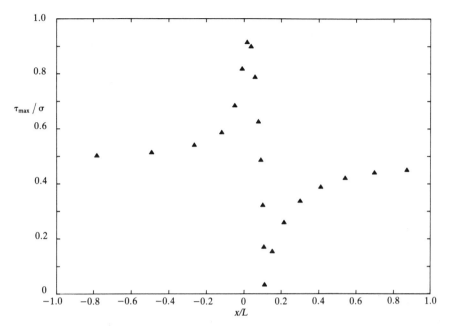

Fig. 9.25—Variation in the maximum shear stress around the fillet radius $R = 0.1L$ in a plate under tension.

end of the plate and $\tau_{max} = \frac{1}{2}\sigma(d/D)$ at the other is seen to involve a localised intensification in the fillet region.

9.10 HIGHER-ORDER ELEMENTS

We have seen that to obtain an acceptable accuracy using CST elements it is generally necessary to employ a refined mesh, particularly in regions where stress concentrations are present. The rate of convergence of the solution obtained using these elements, as the degree of refinement of the mesh is increased, is of the order of l^2. An alternative to the use of highly refined meshes is to employ *higher-order elements* whose displacement functions $[u]$ contain not only linear but also higher-order polynomial terms.

The simplest example of a higher-order triangular plane element has one node along each edge, usually at the midpoint, in addition to the nodes at the vertices (see Fig. 9.26(a)). The displacement functions are approximated by complete second-order polynomials given by Pascal's triangle (see Fig. 6.2) as

$$u = C_1 + C_2x + C_3y + C_4x^2 + C_5xy + C_6y^2$$

$$v = D_1 + D_2x + D_3y + D_4x^2 + D_5xy + D_6y^2$$

where the 12 polynomial coefficients $C_1, D_1, \ldots$ are expressible in terms of the 12 degrees of freedom of the element $u_i, v_i, \ldots$ in the usual way. With three nodes defining each edge of the element, along which the displacements vary

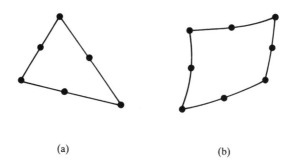

(a)   (b)

Fig. 9.26—Higher-order elements: (a) six-node triangle; (b) eight-node isoparametric quadrilateral.

quadratically, it follows that u and v will be continuous between adjacent elements. The presence of the same constant and linear terms as for the CST element ensures that constant-strain states and rigid-body displacements can be modelled. The element has strains which are linear functions of x and y and is referred to as a *linear-strain triangle* (LST).

Other higher-order triangular plane elements are obtained by approximating u and v as complete polynomials of third and higher order, and by adding additional nodes along the edges of the element. A similar family of quadrilateral elements can also be derived.

A further refinement, in which elements are allowed to have *curved sides* (see Fig. 9.26(b)), involves the use of the same interpolation functions to define the element shape as are used to define the displacement functions. Such *isoparametric elements* have achieved widespread popularity because of their versatility and efficiency, particularly where complicated geometries are to be modelled.

The use of higher-order elements will generally yield a more accurate solution for a mesh of a given degree of refinement. However, a substantial increase in the computing time required results from the need to employ *numerical integration* in the evaluation of $[K^e]$ whose semi-bandwidth is significantly greater than for CST elements.

PROBLEMS

9.1   A vertical line load of intensity P per unit thickness acts on the horizontal boundary of a semi-infinite elastic plate (see Fig. 9.10) in which plane stress conditions obtain. By integrating the strain–displacement equations, and using the boundary condition $u = 0$ at $r = h$ on the plane of symmetry $\theta = 0$, show that the vertical displacement of the free surface $\theta = \pm\pi/2$ is given by

$$v = \pm\frac{P}{\pi E}\left[2\log\left(\frac{r}{h}\right) + (1 + v)\right]$$

9.2   The line load on the semi-infinite plate in problem 9.1 is replaced by a uniform compressive stress σ acting over a length $2a$ of the surface. Using the solution for the radial stress state corresponding to the line load problem, together with the principle of superposition, show that the maximum shear stress at point A is given by

$$\tau_{\max} = \frac{\sigma}{\pi} \sin(\theta_2 - \theta_1)$$

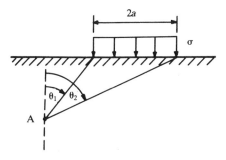

9.3 Show that, for conditions of plane stress, the principal stresses close to the tip of a crack in an elastic plate (see Fig. 3.7) are given by

$$\sigma_{1,2} = \frac{K_I}{\sqrt{2\pi r}} \cos\left(\frac{\theta}{2}\right)\left[1 \pm \sin\left(\frac{\theta}{2}\right)\right], \qquad \sigma_3 = 0$$

If the material yields according to von Mises' criterion, and assuming that the presence of a small plastic zone around the crack tip does not affect the solution for the stresses, show that the extent of the plastic zone is defined by the curve

$$r = \frac{K_I^2}{2\pi\sigma_Y^2}\cos^2\left(\frac{\theta}{2}\right)\left(\frac{5}{2} - \frac{3}{2}\cos\theta\right)$$

9.4 Thermal effects are to be included in the analysis of a plane problem for a solid with a coefficient of thermal expansion α and a temperature $T = T(x, y)$. Show that the constitutive equations are given by

$$[\sigma] = [D]\,([e] - [e_T])$$

where $[e_T] = [\alpha^* T \quad \alpha^* T \quad 0]^T$ is the column matrix of thermal strains in which $\alpha^* = \alpha$ for plane stress and $\alpha^* = \alpha(1 + \nu)$ for plane strain.

Show that, in a finite element formulation of the problem, it is necessary to include an additional term in the equivalent nodal point forces $[F^e]$, defined in equation (6.14), given by

$$\int_{V_e} [B]^T[D][e_T]\,dV.$$

9.5 The straight edge of a plane stress element is parallel to the x axis and is defined by the nodes 1, 2 and 3. The components of displacement u, v along the edge are quadratic functions of x. Using virtual work, show that the equivalent set of nodal point forces corresponding to a prescribed tensile stress $\bar{\sigma}$ which varies quadratically along the edge is given by

$$
\begin{bmatrix} Q_1^e \\ Q_2^e \\ Q_3^e \end{bmatrix} = \frac{l}{30} \begin{bmatrix} 4 & 2 & -1 \\ 2 & 16 & 2 \\ -1 & 2 & 4 \end{bmatrix} \begin{bmatrix} \bar{\sigma}_1 \\ \bar{\sigma}_2 \\ \bar{\sigma}_3 \end{bmatrix}
$$

where $\bar{\sigma}_1$, $\bar{\sigma}_2$ and $\bar{\sigma}_3$ are the nodal point values of σ.

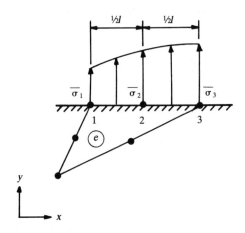

Chapter 10

Axisymmetric Elasticity Problems

10.1 INTRODUCTION

The analysis of stresses in a three-dimensional elastic solid, whose geometry and external loading are both *axisymmetric*, can be achieved using an exact two-dimensional theory.

For convenience, we employ a cylindrical coordinate system in which z is the axis of symmetry (see Fig. 10.1). Any radial rz plane is a plane of symmetry and carries no shear stresses, whilst points on this plane can only undergo displacements in the plane itself (see section 4.6). It follows that

$$\tau_{\theta z} = \tau_{\theta r} = 0 \quad \text{and} \quad v = 0 \tag{10.1}$$

everywhere, that the remaining displacements and stresses are all independent of θ and that the hoop stress $\sigma_{\theta\theta}$ is a principal stress. The strains are given by equations (3.2) rewritten for cylindrical coordinates as

$$e_{rr} = \frac{1}{E} \left[\sigma_{rr} - v(\sigma_{\theta\theta} + \sigma_{zz}) \right], \qquad \gamma_{rz} = \frac{\tau_{rz}}{G}$$

$$e_{\theta\theta} = \frac{1}{E} \left[\sigma_{\theta\theta} - v(\sigma_{zz} + \sigma_{rr}) \right] \tag{10.2}$$

$$e_{zz} = \frac{1}{E} \left[\sigma_{zz} - v(\sigma_{rr} + \sigma_{\theta\theta}) \right]$$

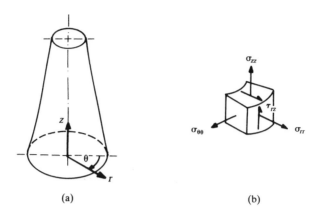

Fig. 10.1—(a) Geometry and (b) state of stress at a point for an axisymmetric problem.

and the strain–displacement relationships (2.28) reduce to

$$e_{rr} = \frac{\partial u}{\partial r}, \qquad e_{\theta\theta} = \frac{u}{r}, \qquad e_{zz} = \frac{\partial w}{\partial z}, \qquad \gamma_{rz} = \frac{\partial u}{\partial z} + \frac{\partial w}{\partial r} \qquad (10.3)$$

The second of the three equations of stress equilibrium (1.30) is automatically satisfied, whilst the remaining equations reduce to

$$\frac{\partial \sigma_{rr}}{\partial r} + \frac{\partial \tau_{rz}}{\partial z} + \left(\frac{\sigma_{rr} - \sigma_{\theta\theta}}{r} \right) + b_r = 0$$

$$\hspace{10cm} (10.4)$$

$$\frac{\partial \tau_{rz}}{\partial r} + \frac{\partial \sigma_{zz}}{\partial z} + \frac{\tau_{rz}}{r} + b_z = 0$$

The stress boundary conditions (1.31) become

$$t_r = \sigma_{rr} l + \tau_{rz} m = \bar{t}_r$$

$$t_z = \tau_{rz} l + \sigma_{zz} m = \bar{t}_z \qquad (10.5)$$

10.2 PLANE AXISYMMETRIC PROBLEMS

The axisymmetric problem for a solid of *uniform* cross-section, where the requirements of plane analysis are also satisfied, is described by a one-dimensional theory. Various problems involving circular discs and cylinders fall into this category and have straightforward analytical solutions.

For plane stress (see section 9.2) the equations of axisymmetric theory are simplifed using the equivalent form of (9.2) for cylindrical coordinates given by

$$\sigma_{zz} = \tau_{zr} = \tau_{z\theta} = 0 \qquad (10.6)$$

It follows that σ_{rr}, $\sigma_{\theta\theta}$ and σ_{zz} are the three principal stresses. The axial body force b_z must be zero, and the radial component b_r must be independent of z. For a solid having a constant angular velocity ω about the z axis, $b_r = \rho\omega^2 r$ and, using (9.9), we can write

$$\sigma_{rr} = \Omega + \frac{1}{r}\frac{d\Phi}{dr}, \qquad \sigma_{\theta\theta} = \Omega + \frac{d^2\Phi}{dr^2} \qquad (10.7)$$

where

$$\frac{d\Omega}{dr} = -b_r = -\rho\omega^2 r$$

and

$$\nabla^2 = \frac{1}{r}\frac{d}{dr}\left(r\frac{d}{dr}\right)$$

The governing equation (9.8) for the stress function now reduces to

$$\frac{1}{r}\frac{d}{dr}\left[r\frac{d}{dr}\left\{\frac{1}{r}\frac{d}{dr}\left(r\frac{d\Phi}{dr}\right)\right\}\right] + (1-v)\frac{1}{r}\frac{d}{dr}\left(r\frac{d\Omega}{dr}\right) = 0$$

which integrates to

$$\Phi = \tfrac{1}{2}Ar^2 + Br^2\log r - C\log r + D + \frac{1-v}{32}\rho\omega^2 r^4 \qquad (10.8)$$

where A, B, C and D are constants. It can be shown that the displacements will be *single valued* only if $B = 0$. The corresponding stresses are, therefore, given by (10.7) as

$$\sigma_{rr} = A - \frac{B}{r^2} - \frac{3+v}{8}\rho\omega^2 r^2$$

$$\qquad (10.9)$$

$$\sigma_{\theta\theta} = A + \frac{B}{r^2} - \frac{1+3v}{8}\rho\omega^2 r^2$$

and the solution for the radial displacement $u = re_{\theta\theta}$ is given by

$$u = \frac{r}{E}(\sigma_{\theta\theta} - v\sigma_{rr})$$

$$\qquad (10.10)$$

$$= \frac{r}{E}\left[A(1-v) + \frac{B}{r^2}(1+v) - \frac{1-v^2}{8}\rho\omega^2 r^2\right]$$

The solution for the plane strain problem (see section 9.3) is found by replacing the elastic constants E and v in (10.9) and (10.10) with E^* and v^* defined in (9.13).

10.3 FINITE ELEMENT FORMULATION

In axisymmetric problems the displacement field is defined by the two components $u = u(r, z)$ and $w = w(r, z)$. The simplest finite element used to analyse these problems is a *ring* of constant triangular cross-section in the rz plane (see Fig. 10.2). The nodal points i, j and k at the vertices are in reality *nodal circles* around which the nodal point quantities remain constant. There is a close similarity between this element and the CST element used for planar analysis in the last chapter. The main difference between the two elements results from the presence of the extra hoop stress and strain components in axisymmetric analysis.

10.3.1 Choice of element displacement functions

The polynomial displacement functions for the element which satisfy all the criteria of section 6.7 are given by

$$u = C_1 + C_2 r + C_3 z$$
$$w = D_1 + D_2 r + D_3 z$$

(10.11)

where the coefficients $C_1, D_1, \ldots$ are constants. Using (10.3), we obtain the corresponding element strains as

$$[e] = \begin{bmatrix} e_{rr} \\ e_{zz} \\ \gamma_{rz} \\ e_{\theta\theta} \end{bmatrix} = \begin{bmatrix} \dfrac{\partial u}{\partial r} \\[2mm] \dfrac{\partial w}{\partial z} \\[2mm] \dfrac{\partial u}{\partial z} + \dfrac{\partial w}{\partial r} \\[2mm] \dfrac{u}{r} \end{bmatrix} = \begin{bmatrix} C_2 \\[2mm] D_3 \\[2mm] C_3 + D_2 \\[2mm] \dfrac{C_1}{r} + C_2 + \dfrac{C_3 z}{r} \end{bmatrix}$$

(10.12)

which are all constant, except the circumferential strain $e_{\theta\theta}$.

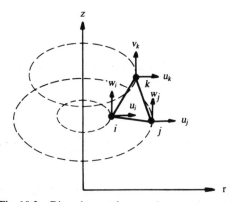

Fig. 10.2—Ring element for an axisymmetric problem.

The only possible constant-strain state for an axisymmetric solid is when u is proportional to r. Such a state is correctly modelled by the functions u and w when $C_1 = C_3 = 0$. Any radial displacement u gives rise to a hoop strain $e_{\theta\theta}$ and, therefore, the only rigid-body displacement possible for an axisymmetric body is a uniform axial displacement w. This is modelled by the functions u and w when $C_1 = C_2 = C_3 = 0$ and $D_2 = D_3 = 0$. To prevent the rigid-body displacement of an assemblage of elements, at least one node must be constrained in the z direction.

10.3.2 Element stiffness matrix

The six polynomial coefficients in (10.11) are evaluated in terms of the nodal displacements $[\delta^e] = [u_i \;\; w_i \;\; u_j \;\; w_j \;\; u_k \;\; w_k]^T$, and by analogy with (9.17) we can write

$$\begin{bmatrix} u \\ w \end{bmatrix} = \begin{bmatrix} N_i & 0 & N_j & 0 & N_k & 0 \\ 0 & N_i & 0 & N_j & 0 & N_k \end{bmatrix} [\delta^e] \tag{10.13}$$

or

$$[u] = [N][\delta^e]$$

where the shape functions are defined by (9.18) with x and y replaced by r and z. By substituting (10.13) into (10.12), we obtain the corresponding element strains as

$$[e] = \frac{1}{2A} \begin{bmatrix} b_i & 0 & b_j & 0 & b_k & 0 \\ 0 & c_i & 0 & c_j & 0 & c_k \\ c_i & b_i & c_j & b_j & c_k & b_k \\ \dfrac{a_i}{r} + b_i + \dfrac{c_i z}{r} & 0 & \dfrac{a_j}{r} + b_j + \dfrac{c_j z}{r} & 0 & \dfrac{a_k}{r} + b_k + \dfrac{c_k z}{r} & 0 \end{bmatrix} [\delta^e]$$

$$= [B][\delta^e]$$

This matrix $[B]$ has one more row than the corresponding matrix for plane problems defined in (9.21) and contains terms which vary over the area of the element.

Generalised Hooke's law (10.2) for an axisymmetric problem has a solution for the stresses given by

$$\begin{bmatrix} \sigma_{rr} \\ \sigma_{zz} \\ \tau_{rz} \\ \sigma_{\theta\theta} \end{bmatrix} = \frac{E}{(1+v)(1-2v)} \begin{bmatrix} 1-v & v & 0 & v \\ v & 1-v & 0 & v \\ 0 & 0 & \frac{1}{2}(1-2v) & 0 \\ v & v & 0 & 1-v \end{bmatrix} \begin{bmatrix} e_{rr} \\ e_{zz} \\ \gamma_{rz} \\ e_{\theta\theta} \end{bmatrix}$$

or

$$[\sigma] = [D][e] \tag{10.15}$$

In the volume integral of (6.10) used to define the element stiffness matrix the differential volume is given by $dV = 2\pi r \, dA$ and therefore $[K^e]$ can be written as the area integral

$$[K^e] = 2\pi \int [B]^T[D][B]r \, dA \tag{10.16}$$

The elements of $[B]$ are no longer constants, as they are in the case of the plane element. Although the integral in (10.16) can be evaluated analytically, the expressions involved are lengthy and involve terms which require special treatment if any of the nodes lie on the axis of symmetry $r = 0$. In practice, it is usually more convenient to approximate the integral numerically by evaluating the integrand at the centroid of the element located at

$$r_c = \tfrac{1}{3}(r_i + r_j + r_k), \qquad z_c = \tfrac{1}{3}(z_i + z_j + z_k) \tag{10.17}$$

By this means we can write (10.16) as

$$[K^e] = 2\pi r_c A \, [B_c]^T[D][B_c] \tag{10.18}$$

where $[B_c]$ denotes the matrix $[B]$ evaluated at the centroid. It should be noted that the approximation deteriorates close to the axis of symmetry where the elements of $[B]$ vary most rapidly.

10.3.3 Stress boundary conditions

The procedure for calculating the equivalent set of nodal point forces for an element whose surface (see Fig. 10.3) carries the linearly varying prescribed tractions

$$[t] = \begin{bmatrix} \bar{t}_r \\ \bar{t}_z \end{bmatrix} = \begin{bmatrix} \sigma_\theta \cos\theta - \tau_\theta \sin\theta \\ \sigma_\theta \sin\theta + \tau_\theta \cos\theta \end{bmatrix} \tag{10.19}$$

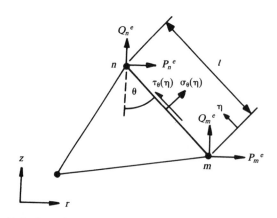

Fig. 10.3—Prescribed boundary stresses for an axisymmetric element.

is similar to that for the plane element (see section 9.4.3). The two elements have identical shape functions, and, by substituting for the differential surface area in (9.24) as $dS = 2\pi r \, d\eta$, we obtain, in place of (9.28), the result

$$[F^e_{mn}] = 2\pi l\left(\int_0^l [N_{mn}]^T[N_{mn}](N_m r_m + N_n r_n)\, d\eta\right)[\bar{t}_{mn}] \tag{10.20}$$

which, after integration, becomes

$$\begin{bmatrix} P^e_m \\ Q^e_m \\ P^e_n \\ Q^e_n \end{bmatrix} = \frac{\pi l}{6}\begin{bmatrix} 3r_m + r_n & 0 & r_m + r_n & 0 \\ 0 & 3r_m + r_n & 0 & r_m + r_n \\ r_m + r_n & 0 & r_m + 3r_n & 0 \\ 0 & r_m + r_n & 0 & r_m + 3r_n \end{bmatrix}\begin{bmatrix} \bar{t}_{r_m} \\ \bar{t}_{z_m} \\ \bar{t}_{r_n} \\ \bar{t}_{z_n} \end{bmatrix} \tag{10.21}$$

10.3.4 Body forces

The equivalent set of nodal point forces for an element subjected to axisymmetric body forces are found by substituting $dV = 2\pi r \, dA$ into (6.11) to give

$$[F^e] = 2\pi \int [N]^T[b] r \, dA \tag{10.22}$$

For body forces which are *constant*, we can approximate this area integral as

$$[F^e] = 2\pi r_c \int [N_i b_r \quad N_i b_z \quad N_j b_r \quad N_j b_z \quad N_k b_r \quad N_k b_z]\, dA$$

and, after substituting for the area integrals $\int N_i \, dA, \ldots$ from (9.31), we obtain

$$[F_e] = \tfrac{2}{3}\pi r_c A[b_i \quad b_j \quad b_i \quad b_j \quad b_i \quad b_j]^T \tag{10.23}$$

If the body forces correspond to the rotation of a solid of density ρ about the z axis with an angular velocity ω, then

$$[b] = \begin{bmatrix} b_r \\ b_z \end{bmatrix} = \begin{bmatrix} \rho\omega^2 r \\ 0 \end{bmatrix} \tag{10.24}$$

By evaluating b_r at the centroid $r = r_c$, and substituting this value into (10.23), we arrive at the approximate set of nodal point forces

$$[F^e] = \tfrac{2}{3}\pi r_c^2 A \rho \omega^2[1 \quad 0 \quad 1 \quad 0 \quad 1 \quad 0]^T \tag{10.25}$$

10.3.5 Element stresses

The element stresses are obtained using (9.33) with $[B]$ replaced by its centroidal value $[B_c]$ to give

$$[\sigma] = [D][B_c][\delta^e] \tag{10.26}$$

10.3.6 Use of FIESTA2

To select the appropriate element subroutine ELEM4 in FIESTA2, we put IETYPE=4 in the data file specifying the finite element model (see Fig. 7.2). The automatic mesh generation facility (see section 7.3) is available if required. For this element the array XY stores the nodal coordinates in the rz plane.

Subroutine ELEM4 calculates the element stiffness array SE using (10.18), and the array FE of nodal point forces corresponding to the combined effect of linearly varying boundary stresses σ_θ, τ_θ and a rotational body force b_r using (10.21) and (10.25), respectively. The array of element stresses SS is calculated in subroutine STR4 using (10.26).

The material data for the problem are specified in the manner described in section 7.2.2. The NOPROP=3 properties for each property set ISET are stored in the array EDAT as follows:

EDAT(ISET,1) = E,   EDAT(ISET,2) = v,   EDAT(ISET,3) = $\rho\omega^2$

Stress boundary conditions are specified in the manner described in section 7.2.3. At least one node must be constrained in the z direction to prevent $[K]$ from being a singular matrix. Failure to do this will result in the error message 'ILL-CONDITIONED EQUATIONS' and the execution of FIESTA2 will be terminated. The only type of force boundary condition that can be specified for this element corresponds to uniformly distributed forces which act around nodal circles.

10.4 CYLINDRICAL PRESSURE VESSELS

10.4.1 Analytical solutions

The stresses in a *long* thick-walled cylindrical vessel of circular cross-section, and subjected to an internal pressure p (see Fig. 10.4), can be found using plane axisymmetric theory (see section 10.2). Away from the end faces of the vessel, it can be assumed that plane cross-sections remain plane during the deformation and that the axial strain e_{zz} is constant (see section 9.3).

Since the plane stress solution for the stresses σ_{rr} and $\sigma_{\theta\theta}$, given by (10.9) with $\omega = 0$, is independent of the elastic constants, it must also be the plane strain solution. The solution which satisfies the boundary conditions $\sigma_{rr} = -p$ at $r = a$ and $\sigma_{rr} = 0$ at $r = b$ is given by *Lame's equations*

$$\sigma_{rr} = \frac{p}{K^2 - 1}\left(1 - \frac{b^2}{r^2}\right), \qquad \sigma_{\theta\theta} = \frac{p}{K^2 - 1}\left(1 + \frac{b^2}{r^2}\right) \qquad (10.27)$$

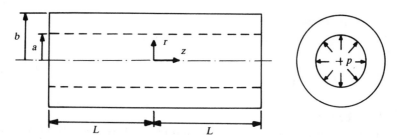

Fig. 10.4—Cylindrical pressure vessel.

where $K = b/a$. The solution for the axial stress σ_{zz} will depend upon the type of end conditions prescribed.

If the rigid end constraints necessary to maintain the plane strain condition $e_{zz} = 0$ are present (see Fig. 10.5(a)), the axial stress is

$$\sigma_{zz} = \nu(\sigma_{rr} + \sigma_{\theta\theta}) = \frac{2\nu p}{K^2 - 1} \tag{10.28}$$

and the radial displacement is given by the second of equations (10.3) as

$$u = re_{\theta\theta}$$
$$= \frac{r}{E}[\sigma_{\theta\theta} - \nu(\sigma_{zz} + \sigma_{rr})] \tag{10.29}$$
$$= \frac{p(1 + \nu)}{E(K^2 - 1)}\left[(1 - 2\nu)r + \frac{b^2}{r^2}\right]$$

When the internal pressure is maintained by opposed pistons (see Fig. 10.5(b), the resultant axial force on any cross-section of the vessel is zero. With $\bar{\sigma}_{zz} = \nu(\sigma_{rr} + \sigma_{\theta\theta})$ in (9.11), it follows that the *plane stress* condition $\sigma_{zz} = 0$ is satisfied in this case, and

$$u = \frac{p}{E(K^2 - 1)}\left[(1 - \nu)r + (1 + \nu)\frac{b^2}{r}\right] \tag{10.30}$$

Finally we have the case in which any cross-section of the vessel carries the tensile load caused by the internal pressure acting over end closures (see Fig. 10.5(c)) such that

$$\int \sigma_{zz}\, dA = p\pi a^2$$

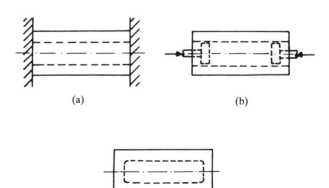

(a)   (b)

(c)

Fig. 10.5—Pressure vessel end conditions: (a) rigid; (b) free; (c) closed.

The axial strain corresponding to this condition is

$$e_{zz} = \frac{p(1 - 2v)}{E(K^2 - 1)}$$

the axial stress is

$$\sigma_{zz} = \frac{p}{K^2 - 1} \tag{10.31}$$

and the corresponding radial displacement is

$$u = \frac{p}{E(K^2 - 1)}\left[(1 - 2v)r + (1 + v)\frac{b^2}{r}\right] \tag{10.32}$$

For all three types of prescribed end condition, $\sigma_{\theta\theta}$ is the maximum, and σ_{rr} the minimum principal stress. According to (1.23) the maximum shear stress at any point is therefore

$$\tau_{max} = \tfrac{1}{2}(\sigma_{max} - \sigma_{min}) = \tfrac{1}{2}(\sigma_{\theta\theta} - \sigma_{rr}) \tag{10.33}$$

10.4.2 Finite element solution

For a long pressure vessel with end closures, shear stresses τ_{rz} are present in the regions adjacent to the ends. To analyse the stresses in these regions, it is no longer possible to assume that e_{zz} is a constant. In a short vessel, it is unlikely that plane theory will be valid even away from the ends. In either case, we must formulate the problem using two-dimensional axisymmetric theory.

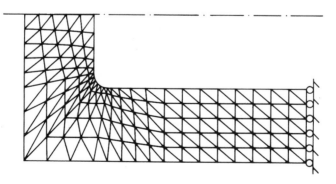

Fig. 10.6—Finite element mesh for a thick cylinder with closed ends.

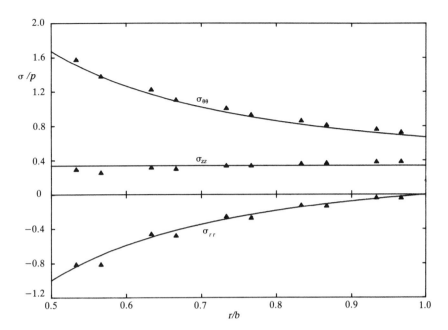

Fig. 10.7—Radial variation in the principal stresses on the cross-section plane of symmetry in
a thick cylinder: ▲, finite element results; ——, plane solutions (10.27), (10.31).

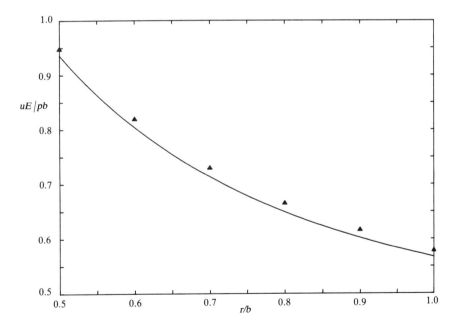

Fig. 10.8—Radial variation in the radial displacement u on the cross-section plane of
symmetry in a thick cylinder: ▲, finite element results; ——, plane solution (10.32).

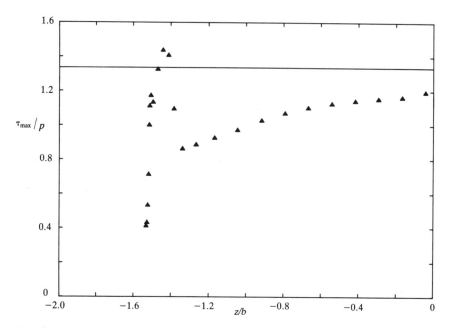

Fig. 10.9—Axial variation in the maximum shear stress τ_{max} along the inner surface of a thick cyclinder: ▲, finite element results; ——, plane solution (10.33).

As an example of the application of the finite element method to this problem, let us consider a vessel with end closures of thickness equal to the wall thickness; $a = b/2$ and $L = 2b$. Because of symmetry, only one-quarter of the diametral cross-section need be modelled (see Fig. 10.6).

Fig. 10.7 shows the radial variations in the three principal stresses on the plane of symmetry $z = 0$ to be in close agreement with the plane solution given by equations (10.27) and (10.31). The radial displacement u on this plane (see Fig. 10.8) is somewhat greater than that given by the plane solution (10.32).

As the junction between the cylinder and the end closures is approached, the radial displacements become progressively less because of the constraint provided by the end closures. It is also found that plane cross-sections of the cylinder become increasingly warped in this region as a result of the presence of the shear stress τ_{rz}.

An indication of the effect of the end closures on the stresses is obtained by examining the axial variation in the maximum shear stress τ_{max} in the elements bordering the inner surface of the cylinder (see Fig. 10.9). The stress undergoes a pronounced fluctuation at the junction with the end closures and the peak value is significantly higher than that predicted by plane theory.

10.5 SPHERICAL PRESSURE VESSEL

To analyse the stresses in a thick-walled spherical pressure vessel (see Fig 10.10), it is convenient to employ *spherical coordinates* in which the position of a point on

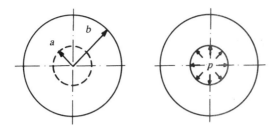

Fig. 10.10—Spherical pressure vessel.

any diametral plane, θ = constant, is specified by the polar coordinates R and ϕ (see Fig. 10.11). The state of stress is defined by the three principal stresses σ_{RR}, $\sigma_{\theta\theta}$ and $\sigma_{\phi\phi} = \sigma_{\theta\theta}$, and the radial displacement u is the only nonzero displacement. When the vessel is subjected to an internal pressure p, the boundary conditions are $\sigma_{RR} = -p$ at $R = a$ and $\sigma_{RR} = 0$ at $R = b$.

The problem is described by a one-dimensional theory, similar to that for cylindrical pressure vessels, in which the stresses and displacements are functions of only the radial coordinate R. Either by solving the problem in terms of u, or by employing a stress function, we obtain

$$\sigma_{RR} = \frac{p}{K^3 - 1}\left(1 - \frac{b^3}{r^3}\right), \qquad \sigma_{\theta\theta} = \frac{p}{K^3 - 1}\left(1 + \frac{b^3}{2r^3}\right) \qquad (10.34)$$

and

$$u = \frac{p}{E(K^3 - 1)}\left[(1 - 2v)r + \frac{1}{2}(1 + v)\frac{b^3}{r^2}\right] \qquad (10.35)$$

where $K = b/a$.

A spherical vessel for which $a = b/2$ is modelled using the finite element mesh shown in Fig. 10.12 and the results obtained for the stresses (see Fig. 10.13) and

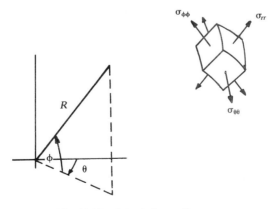

Fig. 10.11—Spherical coordinates.

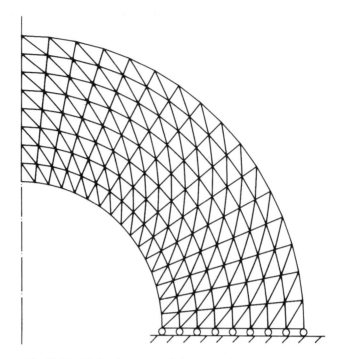

Fig. 10.12—Finite element mesh for a spherical pressure vessel.

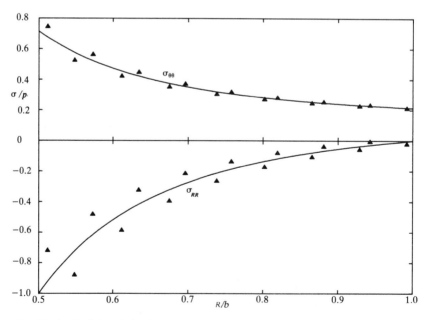

Fig. 10.13—Radial variation in the principal stresses σ_{RR} and $\sigma_{\theta\theta}$ in a spherical pressure vessel: ▲, finite element results; ——, exact solutions (10.34).

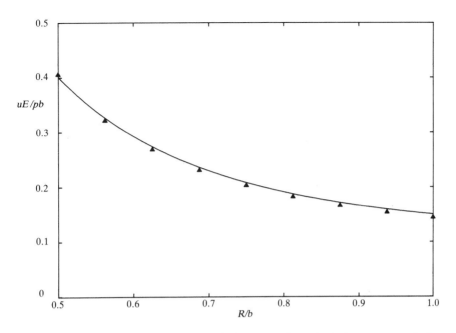

Fig. 10.14—Radial variation in the radial displacement u in a spherical pressure vessel: ▲, finite element results; ——, exact solution (10.35).

the radial displacement (see Fig. 10.14) are in quite close agreement with the exact solutions (10.34) and (10.35).

10.6   CIRCULAR PLATE

We consider a *thin* circular plate of thickness h and radius b and with a central circular hole of radius a (see Fig. 10.15). The plate is simply supported around its outer edge and carries a uniform lateral pressure p on its upper surface. The analytical solution for the maximum lateral deflection δ at the inner radius $r = a$,

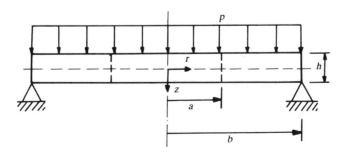

Fig. 10.15—Simply supported circular plate subjected to a lateral pressure p.

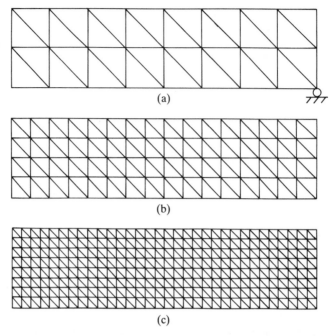

Fig. 10.16—Finite element meshes for a circular plate: (a) mesh 1, 3×9 nodes; (b) mesh 2, 5 × 17 nodes: (c) mesh 3, 9×33 nodes.

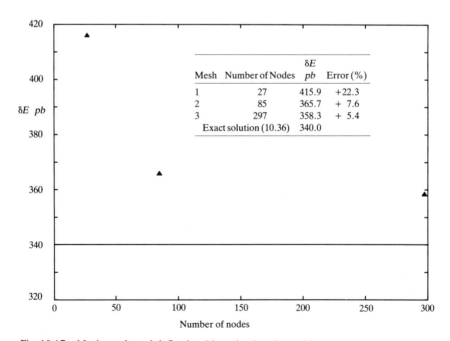

Mesh	Number of Nodes	$\dfrac{\delta E}{pb}$	Error (%)
1	27	415.9	+22.3
2	85	365.7	+ 7.6
3	297	358.3	+ 5.4
Exact solution (10.36)		340.0	

Fig. 10.17—Maximum lateral deflection δ in a circular plate subjected to lateral pressure p: ▲, finite element results; ——, exact solution.

obtained using thin plate theory (see section 4.3), is given in dimensionless form as

$$\frac{\delta E}{pb} = 0.664 \left(\frac{b}{h}\right)^3 \tag{10.36}$$

for a plate with $a = b/2$ and $v = 0.3$.

As the thickness of the plate is increased, so the effect of the shear stress τ_{rz} on the deformation becomes significant and a correction factor must be introduced in (10.36).

The three progessively finer meshes shown in Fig. 10.16 are used to obtain finite element approximations to the problem for $h = b/8$. The results obtained for δ (see Fig. 10.17) converge towards the exact solution (10.36) as the refinement of the mesh is increased.

Using the results for meshes 2 and 3, together with the extrapolation formula (5.27), the improved approximation $\delta E/pb = 355.8$ is obtained.

10.7 ROTATING THIN DISC

The stresses and displacement in a rotating *thin* disc of uniform thickness, inner radius a and outer radius b are given by (10.9) and (10.10). After evaluating the constants A and B using the boundary conditions $\sigma_{rr} = 0$ at $r = a$ and $r = b$, we can write the stresses as

$$\sigma_{rr} = \frac{3 + v}{8} \rho\omega^2 b^2 \left(1 + \frac{a^2}{b^2} - \frac{a^2}{r^2} - \frac{r^2}{b^2}\right)$$

$$\sigma_{\theta\theta} = \frac{3 + v}{8} \rho\omega^2 b^2 \left(1 + \frac{a^2}{b^2} + \frac{a^2}{r^2} - \frac{1 + 3v}{3 + v}\frac{r^2}{b^2}\right) \tag{10.37}$$

and the displacement as

$$u = \frac{r}{E}\frac{3 + v}{8} \rho\omega^2 b^2 \left[\left(1 + \frac{a^2}{b^2}\right)(1 - v) + (1 + v)\frac{a^2}{r^2} - \frac{1 - v^2}{3 + v}\frac{r^2}{b^2}\right] \tag{10.38}$$

The maximum hoop stress occurs at the inner radius $r = a$ where

$$\sigma_{\theta\theta} = \frac{3 + v}{4} \rho\omega^2 b^2 \left(1 + \frac{a^2}{b^2}\frac{1 - v}{3 + v}\right) \tag{10.39}$$

and the maximum radial stress occurs at $r = \sqrt{ab}$ where

$$\sigma_{rr} = \frac{3 + v}{8} \rho\omega^2 (b - a)^2 \tag{10.40}$$

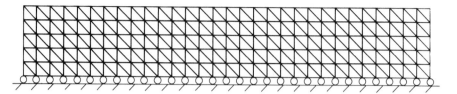

Fig. 10.18—Finite element mesh for a thin rotating disc.

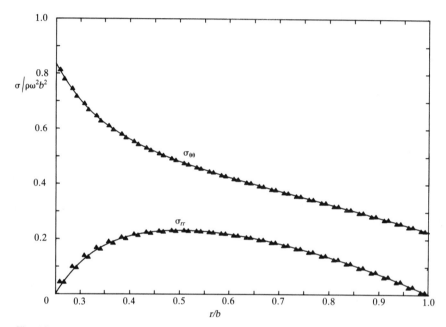

Fig. 10.19—Radial variation in the principal stresses σ_{rr} and $\sigma_{\theta\theta}$ in a thin rotating disc:
▲, finite element results; ——, exact solutions (10.37).

Since $\sigma_{\theta\theta}$ is the maximum, and $\sigma_{zz} = 0$ the minimum principal stress, it follows from (1.23) that the maximum shear stress at any point in the disc is given by

$$\tau_{max} = \tfrac{1}{2}(\sigma_{max} - \sigma_{min}) = \tfrac{1}{2}\sigma_{\theta\theta} \tag{10.41}$$

A finite element solution to the problem is obtained for $a = b/4$ and $\nu = 0.3$ using the mesh shown in Fig. 10.18 and the results shown in Figs 10.19 and 10.20 are in close agreement with the exact solutions (10.37) and (10.38), respectively.

10.8   ROTATING DISC OF VARIABLE THICKNESS

A common component in a variety of rotating machinery is a *rotor* consisting of an integral wheel and hub arrangement with a fillet radius R at their junction to reduce the stress concentration occurring in this region (see Fig. 10.21). The geometry is

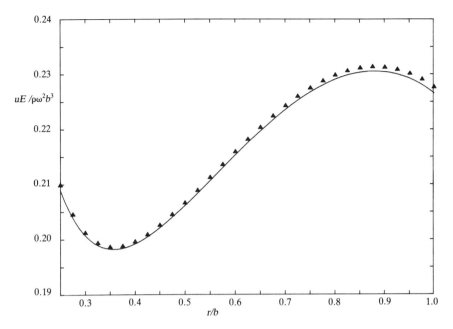

Fig. 10.20—Radial variation in the radial displacement u in a thin rotating disc: ▲, finite element results, ——, exact solution (10.38).

essentially that of a disc of variable thickness, in which the shear stress τ_{rz} is significant. It follows that plane analysis is no longer applicable and that two-dimensional axisymmetric theory must be employed. An analytical solution to the problem is not available.

The finite element mesh used to model a rotor for which $h = b/4$, $R = b/40$, $a = b/4$ and $c = 3b/8$ is shown in Fig. 10.22. The results obtained for the radial variations in the stresses on the plane of symmetry $z = 0$ are compared with the exact solutions for a thin disc with the same inner and outer radii as the rotor in Fig. 10.23. Close to the bore of the rotor the presence of the hub significantly reduces the hoop stress, causes an axial stress to be set up but has little effect on the radial stress. For points remote from the bore the results are in close agreement with the thin disc solution.

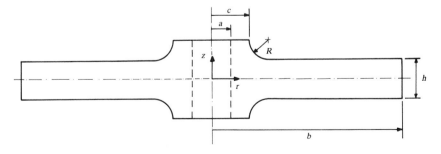

Fig. 10.21—Rotor geometry.

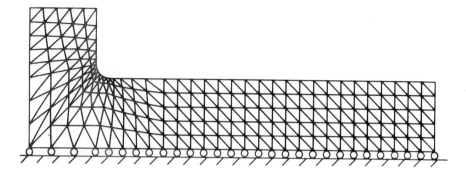

Fig. 10.22—Finite element mesh for a rotor.

The radial variations in stresses in the elements bordering the free surface of the rotor are shown in Fig. 10.24. Exact solutions for a thin disc with inner and outer radii a and b, respectively, and for a thin disc of inner and outer radii a and c, respectively, are included for comparison. Pronounced local stress intensifications occur in the vicinity of the fillet where σ_{zz} has a significant magnitude. Within the hub the hoop stress is substantially higher than that predicted by the thin disc solution for the hub, whilst in the wheel the stresses are quite close to those predicted by the corresponding thin disc solution.

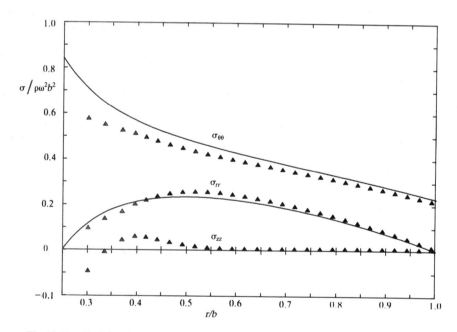

Fig. 10.23—Radial variation in the stresses on the plane of symmetry of a rotor: ▲, finite elements results; —, exact solution (10.37) for a thin disc.

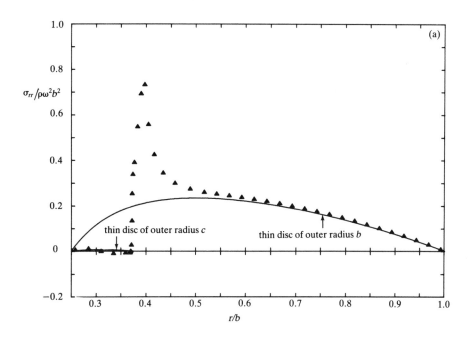

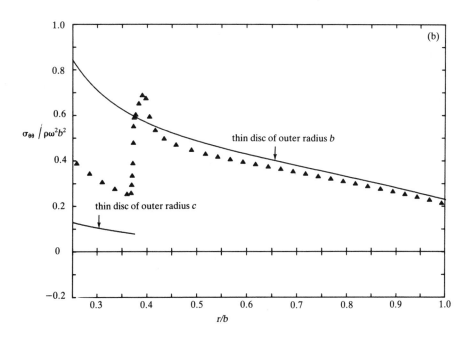

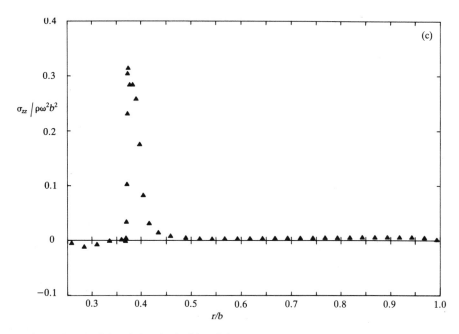

Fig. 10.24—Radial variations in the (a) radial stress σ_{rr}, (b) hoop stress $\sigma_{\theta\theta}$ and (c) axial stress σ_{zz} on the free surface of a rotor.

10.9 ROTATING SHAFT

In a *long* rotating hollow shaft of uniform circular cross-section, it may be assumed that, away from the end faces, plane cross-sections remain plane and that the axial strain e_{zz} is constant (see section 9.3).

The plane strain solution, where $e_{zz} = 0$, is found by replacing the elastic constants E and v in the corresponding plane stress solution by E^* and v^* defined (9.13). After making these modifications, we can write (10.37) and (10.38) as

$$\sigma_{rr} = \frac{1}{8}\frac{3-2v}{1-v}\rho\omega^2 b^2 \left(1 + \frac{a^2}{b^2}\ \frac{a^2}{r^2}\ \frac{r^2}{b^2}\right)$$

$$\sigma_{\theta\theta} = \frac{1}{8}\frac{3-2v}{1-v}\rho\omega^2 b^2 \left(1 + \frac{a^2}{b^2} + \frac{a^2}{r^2} - \frac{1+2v}{3-2v}\frac{r^2}{b^2}\right)$$

(10.42)

and

$$u' = \frac{r(1+v)(3-2v)}{8E(1-v)}\rho\omega^2 b^2 \left[(1-2v)\left(1+\frac{a^2}{b^2}\right) + \frac{a^2}{r^2} - \frac{1-2v}{3-2v}\frac{r^2}{b^2}\right] \quad (10.43)$$

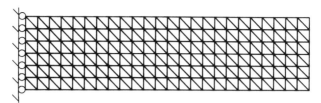

Fig. 10.25—Finite element mesh for a long rotating cyclinder.

The axial stress σ_{zz} remote from the end faces of the shaft defined in (9.11) with

$$\bar{\sigma}_{zz} = \frac{v}{A} \int (\sigma_{rr} + \sigma_{\theta\theta}) \, dA = \tfrac{1}{2} v \rho \omega^2 (a^2 + b^2) \tag{10.44}$$

is given by

$$\sigma_{zz} = v(\sigma_{rr} + \sigma_{\theta\theta}) - \bar{\sigma}_{zz}$$

$$= \frac{v \rho \omega^2 b^2}{4(1 - v)} \left(1 + \frac{a^2}{b^2} - \frac{2r^2}{b^2} \right) \tag{10.45}$$

In the absence of the end constraints necessary to maintain the plane strain condition $e_{zz} = 0$, we must superimpose the effect of a uniform stress $-\bar{\sigma}_{zz}$ for which

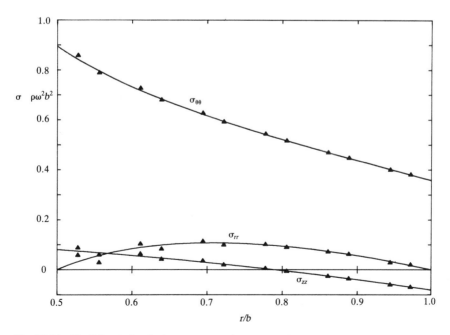

Fig. 10.26—Radial variation in the stresses on the cross-section plane of symmetry of a long rotating cyclinder: ▲, finite element results; ——, exact solutions (10.42), (10.45).

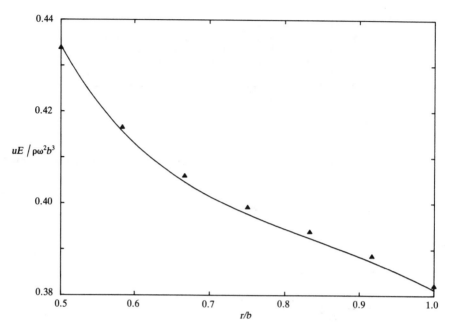

Fig. 10.27—Radial variation in the radial displacement u on the cross-section plane of symmetry of a long rotating cylinder; ▲, finite element results; ——, exact solution (10.47).

the corresponding displacement is

$$u'' = re_{\theta\theta} = \frac{vr\bar{\sigma}_{zz}}{E} = \frac{v^2\rho\omega^2 r}{2E}(a^2 + b^2)$$

(10.46)

It follows that the solution for the radial displacement in the shaft is given by

$$u = u' + u''$$

(10.47)

The results obtained using the finite element mesh shown in Fig. 10.25 for a shaft with $b = L/4$, $a = L/8$ and $v = 0.3$ are in close agreement with the exact solutions (10.42), and (10.45) and (10.47) in the central region of the shaft remote from the end faces (see Figs 10.26 and 10.27). *Barrelling* of the shaft walls is associated with significant warping of plane cross-sections of the shaft in the vicinity of the end faces where plane analysis is no longer applicable.

PROBLEMS

10.1   A thick steel cylinder with closed ends has inner and outer diameters of 240 mm and 300 mm, respectively, and is used as a submersible instrument carrier. Find the greatest depth of water at which the cylinder can be used if there is to be a factor of safety of at least 1.2 against yielding according to Tresca's criterion. The material has a yield stress σ_Y of 240 MN/m^2, and the density of water is 1000 kg/m^3.

(Answer: 3.67 km.)

10.2   A thin steel ring is shrunk onto the outside of a thin solid steel disc. The radius of the interface is 250 mm and the outer radius of the assembly is 350 mm. The contact pressure between the ring and the disc is not to fall below 30 MN/m^2, and the maximum tensile stress in the ring must not exceed 200 MN/m^2. Calculate the maximum speed of rotation of the assembly, and the corresponding stresses at the centre of the solid disc. The density and Poisson's ratio of steel are 7900 kg/m^3 and 0.3, respectively.

(Answers: 3329 rev/min, $\sigma_{rr} = \sigma_{\theta\theta} = -5.3$ MN/m^2.)

10.3   A compound cylinder assembly consists of two concentric long hollow cylinders of the same elastic material, with an interference fit between them. Before assembly the inner cylinder has an outer diameter D_i and the outer cylinder has an inner diameter D_o, where $\delta = D_i - D_o\,(>0)$ denotes the diametral interference. If the inner and outer radii of the compound cylinder after assembly are a and b, respectively, and the radius of the interface between the cylinders is c, show that

$$\delta = \frac{2c}{E}\,\Delta\sigma_{\theta\theta}$$

where $\Delta\sigma_{\theta\theta}$ denotes the discontinuity in the magnitude of the hoop stress at the interface. Hence, deduce the formula for the contact pressure p at the interface given by

$$p = \frac{E\delta}{4c^3}\,\frac{(c^2 - a^2)(b^2 - c^2)}{b^2 - a^2}$$

10.4   Show that the equation of equilibrium in a plane stress axisymmetric problem, for a solid in which the temperature variation is of the form $T = T(r)$, is given by

$$\frac{d^2u}{dr^2} + \frac{1}{r}\frac{du}{dr} - \frac{u}{r^2} = \alpha(1 + v)\frac{dT}{dr}$$

Hence obtain the solution for the stresses

$$\sigma_{rr} = A - \frac{B}{r^2} - \frac{\alpha E}{r^2}\int Tr\,dr$$

$$\sigma_{\theta\theta} = A + \frac{B}{r^2} + \frac{\alpha E}{r^2}\int Tr\,dr - \alpha ET$$

where A and B are constants.

10.5   Derive from first principles the equation of equilibrium in spherical coordinates (R, θ, ϕ) for an internally pressurised thick-walled spherical pressure vessel. Using the strain–displacement relationships $e_{RR} = du/dR$ and $e_{\theta\theta} = u/R$, show that equilibrium can be expressed in terms of the radial displacement u as

$$\frac{d^2u}{dR^2} + \frac{2}{R}\frac{du}{dR} - \frac{2u}{R^2} = 0$$

Hence solve for the stresses σ_{RR} and $\sigma_{\theta\theta}$.

(Answers: $d\sigma_{RR}/dR + (2/R)(\sigma_{RR} - \sigma_{\theta\theta}) = 0$; $A - 2B/R^3$, $A + B/R^3$.)

Chapter 11

Torsion of Elastic Shafts

11.1 INTRODUCTION

The elastic solution for a circular shaft of uniform diameter, subjected to a torsional load T (see Fig. 11.1), is given by the *elementary theory of torsion.* By assuming that plane cross-sections remain plane during the twisting of the shaft, it can be shown that the shear stress τ at a radius r, and the angle α of twist per unit length are related through

$$\frac{T}{J} = \frac{\tau}{r} = G\alpha \tag{11.1}$$

where $J = \int r^2 \, dA$ is the polar second moment of area of the cross-section and G is the elastic shear modulus of the material.

In this chapter, we examine two important torsional problems not covered by the elementary theory. These involve shafts of uniform but noncircular section, and circular shafts of variable diameter.

11.2 NONCIRCULAR SHAFT

Shafts are often of noncircular section, at least over a portion of their length. Examples include splined drive shafts, and shafts with keyways through which the applied torque is transmitted. Such problems are formulated using the two-dimensional theory of St Venant.

Consider a uniform shaft of arbitrary section carrying a torsional load T (see Fig. 11.2). The end face $z = 0$ is fixed, and the total angle of twist of the shaft is assumed to be small. The deformation of every slice Δz of the shaft away from the end faces may be assumed to be the same. It follows that α is constant along the length of the shaft and the angle of twist at any cross-section is given by $\phi = \alpha z$.

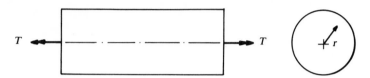

Fig. 11.1—Uniform circular shaft.

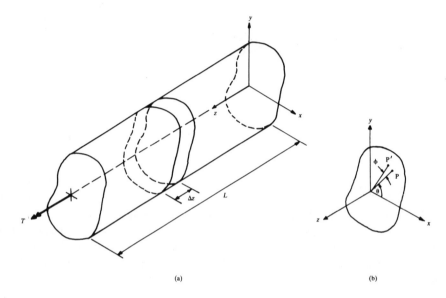

(a)   (b)

Fig. 11.2—(a) Geometry and (b) deformation of a uniform noncircular shaft.

The components of displacement in the xy plane of a point on the cross-section as it moves from P to P′ during the deformation are

$$u = -r\phi \sin\theta = -y\phi = -\alpha yz$$
$$v = r\phi \cos\theta = x\phi = \alpha xz \tag{11.2}$$

The third component w, which describes the *warping* of cross-sections, is independent of z and may be written as

$$w = \alpha\psi(x, y) \tag{11.3}$$

By substituting (11.2) and (11.3) into (2.15), we obtain the strains as

$$e_{xx} = e_{yy} = e_{zz} = \gamma_{xy} = 0$$

$$\gamma_{yz} = \frac{\partial v}{\partial z} + \frac{\partial w}{\partial y} = \alpha\left(\frac{\partial\psi}{\partial y} + x\right)$$

$$\gamma_{zx} = \frac{\partial w}{\partial x} + \frac{\partial u}{\partial z} = \alpha\left(\frac{\partial\psi}{\partial x} - y\right) \tag{11.4}$$

The strain compatibility condition is obtained by eliminating ψ between equations (11.4) to give

$$\frac{\partial \gamma_{yz}}{\partial x} - \frac{\partial \gamma_{zx}}{\partial y} = 2\alpha \qquad (11.5)$$

The corresponding elastic stresses given by (3.4) are

$$\sigma_{xx} = \sigma_{yy} = \sigma_{zz} = \tau_{xy} = 0$$

$$\tau_{yz} = \alpha G\left(\frac{\partial \psi}{\partial y} + x\right), \qquad \tau_{zx} = \alpha G\left(\frac{\partial \psi}{\partial x} - y\right) \qquad (11.6)$$

By solving for the roots of (1.22), we obtain the principal stresses as

$$\sigma_1 = 0 \qquad \text{and} \qquad \sigma_{2,3} = \pm\sqrt{\tau_{yz}^2 + \tau_{zx}^2} \qquad (11.7)$$

and according to (1.23) the maximum shear stress at any point is

$$\tau_{max} = \sqrt{\tau_{yz}^2 + \tau_{zx}^2} \qquad (11.8)$$

Equilibrium (1.25) reduces to the single equation

$$\frac{\partial \tau_{zx}}{\partial x} + \frac{\partial \tau_{yz}}{\partial y} = 0 \qquad (11.9)$$

and the condition $\bar{t}_x = \bar{t}_y = \bar{t}_z = 0$ for the surface of the shaft to be free from external forces is obtained from (1.26) as

$$\tau_{zx}l + \tau_{yz}m = 0 \qquad (11.10)$$

where $l = dy/ds$ and $m = -dx/ds$ (see Fig. 11.3). The torque on any cross-section of the shaft is

$$T = \iint_A (x\tau_{yz} - y\tau_{zx})\, dx\, dy \qquad (11.11)$$

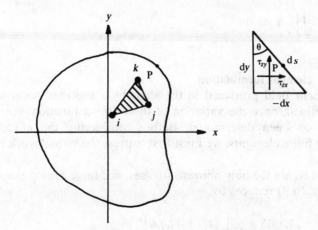

Fig. 11.3—Cross-section of a uniform non-circular shaft.

It is assumed that the solution for the stresses is the same for every section of the shaft, including the end faces. For a long shaft, any discrepancies between the actual distribution of tractions over the ends, and that required by the solution will, according to St Venant's principle (see section 4.4), only be significant near the ends.

11.2.1 Analytical formulation

With the stresses defined in terms of a stress function $\Phi(x, y)$ as

$$\tau_{yz} = -\frac{\partial \Phi}{\partial x}, \qquad \tau_{zx} = \frac{\partial \Phi}{\partial y} \tag{11.12}$$

equilibrium (11.9) is automatically satisfied, and the strains can be expressed as

$$\gamma_{yz} = -\frac{1}{G}\frac{\partial \Phi}{\partial x}, \qquad \gamma_{zx} = \frac{1}{G}\frac{\partial \Phi}{\partial y} \tag{11.13}$$

After substituting (11.13) into (11.5), we obtain the compatibility condition

$$\nabla^2 \Phi = -2G\alpha \tag{11.14}$$

The boundary condition (11.10) can be written in terms of Φ using (11.12) as

$$\frac{\partial \Phi}{\partial y}\frac{dy}{ds} + \frac{\partial \Phi}{\partial x}\frac{dx}{ds} = \frac{d\Phi}{ds} = 0$$

which shows that the stress function must be *constant* along the boundary. In the case of a singly connected section (i.e. one defined by a single closed curve), we take

$$\Phi = 0 \tag{11.15}$$

on the boundary, for convenience.

Using (11.12) and (11.15), we can express the torque defined in (11.11) as

$$T = 2\iint_A \Phi \, dx \, dy \tag{11.16}$$

11.2.2 Finite element formulation

The displacement field produced in the shaft by a specified torsional load T is completely defined, once the variation of the warping function $\psi(x, y)$ over any cross-section has been determined. Before considering the modelling of this function using finite elements, we must first express the virtual work balance (6.9) in terms of ψ.

Since τ_{yz} and τ_{zx} are the only nonzero stresses, and there are no external forces in the z direction, (6.9) reduces to

$$\int (ut_x + _vt_y) \, dS = \frac{1}{G}\int (\tau_{yz}^2 + \tau_{zx}^2) \, dV \tag{11.17}$$

The surface integral on the left-hand side of (11.17) denotes the virtual work $T\alpha L$ done by the applied torque T through the angle of twist αL. Taking the definition (11.11) for T with the stresses expressed in terms of ψ using (11.6), we can write this integral as

$$T\alpha L = \alpha^2 LG \iint_A \left(x \frac{\partial \psi}{\partial y} - y \frac{\partial \psi}{\partial x} + x^2 + y^2 \right) dx\, dy \tag{11.18}$$

The volume integral on the right-hand side of (11.17) has an integrand which is independent of z and can be expressed in terms of ψ, using (11.6), as

$$\frac{1}{G} \int (\tau_{yz}^2 + \tau_{zx}^2)\, dV = \alpha^2 LG \iint_A \left[\left(\frac{\partial \psi}{\partial y} + x \right)^2 + \left(\frac{\partial \psi}{\partial x} - y \right)^2 \right] dx\, dy \tag{11.19}$$

Equating the left- and right-hand sides of the virtual work balance defined in (11.18) and (11.19), respectively, we have

$$\iint_A \left(-x \frac{\partial \psi}{\partial y} + y \frac{\partial \psi}{\partial x} \right) dx\, dy = \iint_A \left[\left(\frac{\partial \psi}{\partial x} \right)^2 + \left(\frac{\partial \psi}{\partial y} \right)^2 \right] dx\, dy \tag{11.20}$$

Consider now the use of triangular finite elements with nodes i, j and k to represent the cross-section of the shaft (see Fig. 11.3). Within a typical element e the warping function is approximated in terms of its nodal point values $[\delta^e] = [\psi_i \ \psi_j \ \psi_k]^T$ by

$$\psi = [N][\delta^e] \tag{11.21}$$

where the linear shape functions $[N]$ are defined in (9.18). After substituting (11.21) into (11.20), we arrive at the set of equilibrium equations for the element given by

$$[F^e] = [K^e][\delta^e]$$

in which the element stiffness matrix is defined by

$$[K^e] = \int_{S_e} \left(\left[\frac{\partial N}{\partial x} \right]^T \left[\frac{\partial N}{\partial x} \right] + \left[\frac{\partial N}{\partial y} \right]^T \left[\frac{\partial N}{\partial y} \right] \right) dS \tag{11.22}$$

and the nodal point forces are

$$[F^e] = \int_{S_e} \left(-x \left[\frac{\partial N}{\partial y} \right]^T + y \left[\frac{\partial N}{\partial x} \right]^T \right) dS \tag{11.23}$$

By differentiating the shape functions in (9.18), we obtain

$$\left[\frac{\partial N}{\partial x} \right] = \frac{1}{2A} [b_i \ \ b_j \ \ b_k], \qquad \left[\frac{\partial N}{\partial y} \right] = \frac{1}{2A} [c_i \ \ c_j \ \ c_k] \tag{11.24}$$

with the aid of which we can evaluate (11.22) and (11.23) as

$$[K^e] = \frac{1}{4A} \begin{bmatrix} b_i^2 + c_i^2 & b_i b_j + c_i c_j & b_i b_k + c_i c_k \\ b_i b_j + c_i c_j & b_j^2 + c_j^2 & b_j b_k + c_j c_k \\ b_i b_k + c_i c_k & b_j b_k + c_j c_k & b_k^2 + c_k^2 \end{bmatrix} \qquad (11.25)$$

and

$$[F^e] = \frac{1}{2A} \begin{bmatrix} -x_c c_i + y_c b_i \\ -x_c c_j + y_c b_j \\ -x_c c_k + y_c b_k \end{bmatrix} \qquad (11.26)$$

where (x_c, y_c) are the coordinates of the element's centroid defined in (10.17) with r and z replaced by x and y.

The set of equilibrium equations for the system of elements can now be assembled in the usual way. The principle of virtual work provides a complete formulation for the problem and it is not necessary to specify boundary conditions for the surface nodes. However, to eliminate rigid-body displacements, we must prescribe a value for ψ, say zero, at one point in the mesh.

The element stresses are obtained from (11.6) as

$$\frac{\tau_{yz}}{\alpha G} = \frac{\partial \psi}{\partial y} + x = \left[\frac{\partial N}{\partial y} \right] [\delta^e] + x$$

and

$$\frac{\tau_{zx}}{\alpha G} = \frac{\partial \psi}{\partial x} - y = \left[\frac{\partial N}{\partial x} \right] [\delta^e] - y$$

and, after substituting for the derivatives of $[N]$ from (11.24), we can write these as

$$\frac{\tau_{yz}}{\alpha G} = \frac{1}{2A} (c_i \psi_i + c_j \psi_j + c_k \psi_k) + x$$
$$\frac{\tau_{zx}}{\alpha G} = \frac{1}{2A} (b_i \psi_i + b_j \psi_j + b_k \psi_k) - y \qquad (11.27)$$

The torsional constant $T/\alpha G$ is obtained from (11.18) as

$$\frac{T}{\alpha G} = \iint_A \left(x \frac{\partial \psi}{\partial y} - y \frac{\partial \psi}{\partial x} \right) dx\, dy + \iint_A (x^2 + y^2)\, dx\, dy$$

and may be approximated by summing over the system of n_e elements representing the cross-section of the shaft to give

$$\frac{T}{\alpha G} = \sum_{e=1}^{n_e} (I_e + J_e) \qquad (11.28)$$

where

$$I_e = \int \left(x \frac{\partial \psi}{\partial y} - y \frac{\partial \psi}{\partial x} \right) dA$$

$$= \int \left(x \left[\frac{\partial N}{\partial y} \right] [\delta^e] - y \left[\frac{\partial N}{\partial x} \right] [\delta^e] \right) dA$$

$$= \tfrac{1}{2}(x_c c_i - y_c b_i)\psi_i + \tfrac{1}{2}(x_c c_j - y_c b_j)\psi_j + \tfrac{1}{2}(x_c c_k - y_c b_k)\psi_k$$

and

$$J_e = \int (x^2 + y^2) \, dA$$

$$= \frac{A}{6} (x_i^2 + x_j^2 + x_k^2 + y_i^2 + y_j^2 + y_k^2 + x_i x_j + x_j x_k + x_k x_i$$
$$+ y_i y_j + y_j y_k + y_k y_i)$$

11.2.3 Use of FIESTA2

To select the appropriate element subroutine ELEM5 in FIESTA2, we put IETYPE=5 in the data file specifying the finite element model (see Fig. 7.2). The automatic mesh generation facility (see section 7.3) is available if required.

Subroutine ELEM5 calculates the element stiffness array SE using (11.25), and the array FE of nodal point forces using (11.26). Subroutine STR5 stores the centroidal values of the element stresses in the array SS using (11.27) with $x = x_c$ and $y = y_c$. The summation for the torsional constant $T/\alpha G$ is performed according to (11.28) and the result is stored in SS(1, 3).

Since $[F^e]$ and $[K^e]$ defined in (11.22) and (11.23), respectively, are independent of the shear modulus G of the element, there are no material data to be specified for the problem. We signify this by putting NOSET=0.

At least one displacement boundary condition $\psi = 0$ must be specified at some point in the mesh to eliminate rigid-body displacements. There are no other boundary conditions to be specified and we therefore put NOPS=0 and NOPF=0 in the data file.

11.2.4 Shaft with square cross-section

For a shaft of square cross-section, of side length $2a$, an analytical solution can be obtained in the form of an infinite series for the stress function. The shear stress is a maximum at the midpoints of the sides where

$$\frac{\tau_{max} a^3}{T} = 0.6010 \tag{11.29}$$

and the dimensionless torsional constant is

$$\frac{T}{\alpha G a^4} = 2.250 \tag{11.30}$$

In a finite element analysis of the problem, we model one quadrant of the cross-section bounded by the planes of symmetry $x = 0$ and $y = 0$ along which we impose the boundary condition $\psi = 0$. The dimensionless torsional constant is evaluated using (11.28) by summing over the elements and multiplying the result by 4. The three successively finer meshes shown in Fig. 11.4 were used, and the results obtained for the maximum shear stress and the torsional constant (see Table 11.1) converge towards the exact solutions (11.29) and (11.30) as the number of nodes are increased.

Table 11.1 Torsional constant and maximum shear stress for a shaft of square cross-section

Number of nodes	$T/\alpha G a^4$	$\tau_{max} a^3/T$
5×5	2.298	0.5426
9×9	2.262	0.5765
17×17	2.252	0.5901
Exact solution	2.250	0.6010

11.2.5 Shaft with an elliptical cross-section

We start by obtaining the analytical solution for a shaft of solid elliptical cross-section (see Fig. 11.5(a)) whose boundary is defined by $x^2/a^2 + y^2/b^2 = 1$. The boundary condition (11.15) is satisfied by taking the solution for the stress function in the form

$$\Phi = C \left(\frac{x^2}{a^2} + \frac{y^2}{b^2} - 1 \right) \tag{11.31}$$

which also satisfies compatibility (11.14) if the constant C is taken as

$$C = -\alpha G \frac{a^2 b^2}{a^2 + b^2}$$

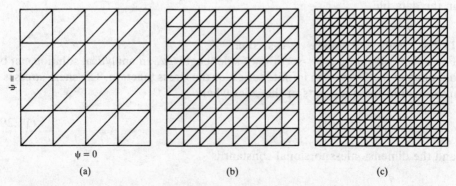

Fig. 11.4—Finite element meshes for one quadrant of a shaft of square cross-section; (a) mesh 1, 5×5 nodes; (b) mesh 2, 9×9 nodes; (c) mesh 3, 17×17 nodes.

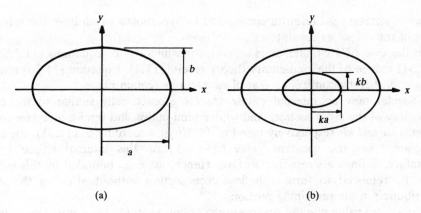

(a)   (b)

Fig. 11.5—Shaft of (a) solid and (b) hollow elliptical cross-section.

After substituting (11.31) into (11.16), we obtain

$$\frac{T}{\alpha G} = -\frac{2a^2b^2}{a^2+b^2}\left(\frac{1}{a^2}\iint x^2 \, dx \, dy + \frac{1}{b^2}\iint y^2 \, dx \, dy - \iint dx \, dy\right)$$

$$= \frac{\pi a^3 b^3}{a^2+b^2} \tag{11.32}$$

and the stresses defined in (11.12) are

$$\frac{\tau_{yz}}{\alpha G} = \frac{2b^2x}{a^2+b^2}, \qquad \frac{\tau_{zx}}{\alpha G} = -\frac{2a^2y}{a^2+b^2} \tag{11.33}$$

The maximum shear stress at a point, given by (11.8) as

$$\frac{\tau_{max}}{\alpha G} = \frac{2}{a^2+b^2}\sqrt{a^4y^2+b^4x^2} \tag{11.34}$$

has a maximum value at the ends $y = \pm b$ of the minor axis of the ellipse given by

$$\frac{\hat{\tau}_{max}}{\alpha G} = \frac{2a^2b}{a^2+b^2} \tag{11.35}$$

The solution for ψ is obtained by first writing (11.6) as

$$\frac{\partial\psi}{\partial y} = \frac{\tau_{yz}}{\alpha G} - x, \qquad \frac{\partial\psi}{\partial x} = \frac{\tau_{zx}}{\alpha G} + y$$

then substituting for the stresses from (11.33) and finally integrating to give

$$\psi = \frac{xy(b^2-a^2)}{a^2+b^2} \tag{11.36}$$

Lines of constant ψ therefore correspond to hyperbolae which have the principal axes of the ellipse as asymptotes.

In the case of a circular cross-section, for which $a = b$, equations (11.32) and (11.34) reduce to the elementary theory results (11.1). Equation (11.36) reduces to $\psi = 0$, thereby confirming that plane sections remain plane.

Consider now an internal ellipse that is geometrically similar to the outer boundary of the cross-section, and whose semi-major and semi-minor axes are of lengths ka and kb respectively (see Fig. 11.5(b)). According to (11.31), the stress function takes the constant value $C(k^2 - 1)$ on this internal ellipse which, therefore, defines a stress-free surface. Hence, material bounded by this surface may be removed to form a hollow cross-section without affecting the stress distribution in the remaining portion.

For a given value of α the stresses in the hollow shaft are the same as those in the solid shaft. The torque will be smaller, however, by an amount equal to the torque carried by the material removed. Instead of (11.32), we now have

$$\frac{T}{\alpha G} = \frac{\pi a^3 b^3}{a^2 + b^2} (1 - k^4) \tag{11.37}$$

In a finite element analysis of a hollow shaft with $b = a/2$ and $k = 0.5$, it is only necessary to model one-quarter of the cross-section bounded by the planes of symmetry $x = 0$ and $y = 0$ (see Fig. 11.6). The dimensionless torsional constant $T/\alpha Ga^4$ has the value 0.294 compared with the exact value of 0.295 given by (11.37). The results obtained for the variation in τ_{max} along the semi-major and semi-minor axes, are compared with the exact solution (11.34) in Fig. 11.7.

11.2.6 Circular shaft with four axial holes

A shaft of circular cross-section, of radius a, has four axial holes each of radius $R = a/5$ whose centres lie on a circle of radius $\rho = 3a/5$ (see Fig. 11.8). The shaft

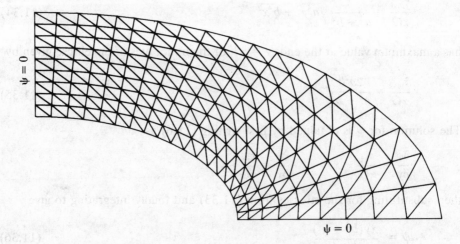

Fig. 11.6—Finite element mesh for a hollow elliptical cross-section.

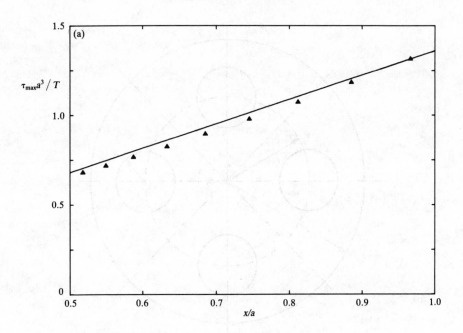

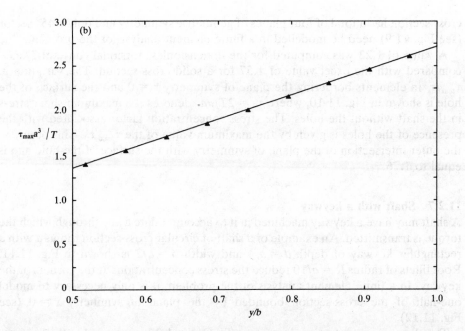

Fig. 11.7—Variation in the maximum shear stress τ_{max} along (a) the major axis and (b) the minor axis of a hollow elliptical cross-section: ▲, finite element results; ——, exact solution (11.34).

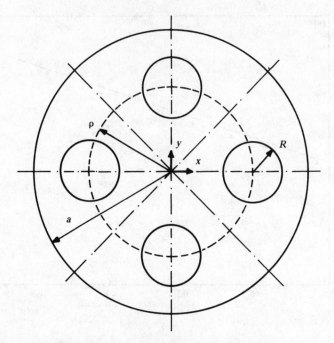

Fig. 11.8—Circular shaft with four axial holes.

cross-section has a total of four planes of geometric symmetry and only a 45° sector (see Fig. 11.9) need be modelled in a finite element analysis of the problem.

A value of 1.23 was computed for the dimensionless torsional constant $T/\alpha Ga^4$ compared with the exact value of 1.57 for a solid cross-section. The variation in τ_{max}/τ_0 in elements bordering the plane of symmetry $x = 0$ and the surface of the hole is shown in Fig. 11.10, where $\tau_0 = 2T/\pi a^3$ denotes the maximum shear stress in the shaft without the holes. The stress concentration factor associated with the presence of the holes is given by the maximum value of the τ_{max}/τ_0. This occurs at the outer intersection of the plane of symmetry with the surface of the hole and is equal to 1.76.

11.2.7 Shaft with a keyway
A shaft may have a keyway machined in it to accommodate a key through which the torque is transmitted. An example of a shaft of circular cross-section radius a with a rectangular keyway of depth $d = a/4$ and width $b = a/2$ is shown in Fig. 11.11. Root fillets of radius $R = a/10$ reduce the stress concentrations at the corners of the keyway. In a finite element analysis of the problem, it is only necessary to model one-half of the cross-section bounded by the plane of symmetry $x = 0$ (see Fig. 11.12).

The dimensionless torsional constant $T/\alpha Ga^4$ is 1.37, compared with the exact value of 1.57 for a circular shaft without a keyway. The variation in τ_{max}/τ_0 in the elements bordering the surface of the keyway is shown in Fig. 11.13, where

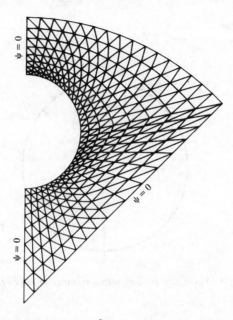

Fig. 11.9—Finite element mesh for a 45° sector of the cross-section of a circular shaft with four axial holes.

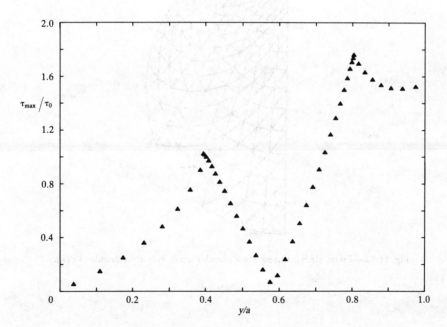

Fig. 11.10—Variation in the maximum shear stress τ_{max} along the boundary of a 45° sector of the cross-section of a circular shaft with four axial holes.

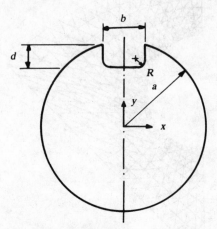

Fig. 11.11—Circular shaft with a rectangular keyway.

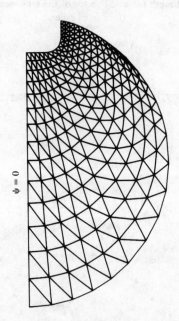

Fig. 11.12—Finite element mesh for a circular shaft with a rectangular keyway.

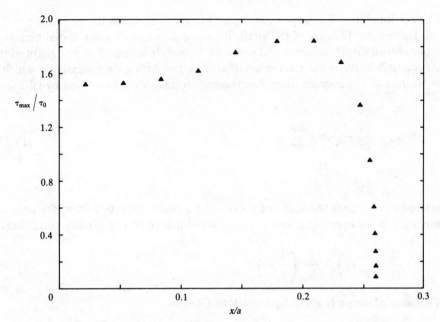

Fig. 11.13—Variation in the maximum shear stress τ_{max} around a rectangular keyway.

$\tau_0 = 2T/\pi a^3$ is the maximum shear stress in the shaft without the keyway. The stress concentration factor, corresponding to the maximum value of the ratio τ_{max}/τ_0 at a point about 15° from the bottom of the keyway on the fillet radius, is equal to 1.84.

11.3 AXISYMMETRIC SHAFT

We turn now to the torsional problem for a circular shaft whose diameter varies rapidly along its length (see Fig. 11.14), as in the case of a stepped shaft with shoulder fillets, or a shaft with a circumferential groove.

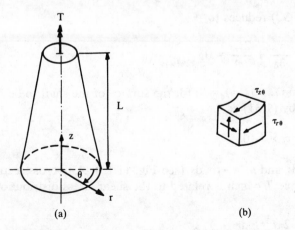

Fig. 11.14—(a) Geometry and (b) state of stress at a point for an axisymmetric shaft.

The torsional load T is applied at the end face $z = L$ whilst the end face $z = 0$ is fixed. During the twisting of the shaft, it can be assumed that the radial and axial components of displacement u and w are both zero. It follows that any point in the shaft can only move in the circumferential direction with a displacement v which is independent of the coordinate θ. The state of strain is therefore given by (2.28) as

$$e_{rr} = e_{\theta\theta} = e_{zz} = \gamma_{rz} = 0$$

$$\gamma_{r\theta} = \frac{\partial v}{\partial r} - \frac{v}{r} = r \frac{\partial \psi}{\partial r} \tag{11.38}$$

$$\gamma_{\theta z} = \frac{\partial v}{\partial z} = r \frac{\partial \psi}{\partial z}$$

where $\psi(r, z) = v/r$ is the angle of rotation of a point distance r from the axis. By eliminating ψ between equations (11.38), we obtain the compatibility condition as

$$\frac{\partial}{\partial r} \left(\frac{1}{r} \gamma_{\theta z} \right) - \frac{\partial}{\partial z} \left(\frac{1}{r} \gamma_{r\theta} \right) = 0 \tag{11.39}$$

The state of stress is given by equations (3.4) as

$$\sigma_{rr} = \sigma_{\theta\theta} = \sigma_{zz} = \tau_{rz} = 0$$

$$\tau_{r\theta} = Gr \frac{\partial \psi}{\partial r}, \qquad \tau_{\theta z} = Gr \frac{\partial \psi}{\partial z} \tag{11.40}$$

for which the principal stresses are the roots of (1.22) given by

$$\sigma_1 = 0, \qquad \sigma_{2,3} = \pm\sqrt{\tau_{r\theta}^2 + \tau_{\theta z}^2} \tag{11.41}$$

and the maximum shear stress at any point is given by (1.23) as

$$\tau_{max} = \sqrt{\tau_{r\theta}^2 + \tau_{\theta z}^2} \tag{11.42}$$

Equilibrium (1.30) reduces to

$$\frac{\partial \tau_{r\theta}}{\partial r} + \frac{\partial \tau_{\theta z}}{\partial z} + \frac{2\tau_{r\theta}}{r} = 0 \tag{11.43}$$

and the condition $\bar{\tau}_r = \bar{\tau}_\theta = \bar{\tau}_z = 0$ for the surface of the shaft to be free of external forces is given by (1.31) as

$$\tau_{r\theta}l + \tau_{\theta z}n = 0 \tag{11.44}$$

where $l = dz/ds$ and $n = -dr/ds$ (see Fig. 11.15). Any cross-section of the shaft carries the torque T which is related to the shear stress distribution $\tau_{\theta z}$ by

$$T = \int 2\pi r^2 \tau_{\theta z} \, dr \tag{11.45}$$

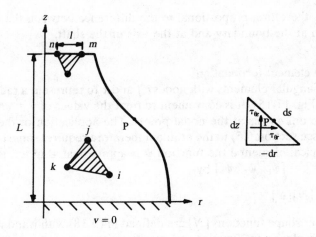

Fig. 11.15—Radial section through an axisymmetric shaft.

11.3.1 Analytical formulation

When the stresses are defined in terms of a stress function $\Phi(r, z)$ as

$$\tau_{r\theta} = - \frac{1}{r^2} \frac{\partial \Phi}{\partial z}, \qquad \tau_{\theta z} = \frac{1}{r^2} \frac{\partial \Phi}{\partial r} \tag{11.46}$$

equilibrium (11.43) is automatically satisfied, and the strains are

$$\gamma_{r\theta} = - \frac{1}{Gr^2} \frac{\partial \Phi}{\partial z}, \qquad \gamma_{\theta z} = \frac{1}{Gr^2} \frac{\partial \Phi}{\partial r} \tag{11.47}$$

Using (11.46), we can write compatibility (11.39) as

$$\frac{\partial^2 \Phi}{\partial r^2} - \frac{3}{r} \frac{\partial \Phi}{\partial r} + \frac{\partial^2 \Phi}{\partial z^2} = 0 \tag{11.48}$$

and the boundary condition (11.44) as

$$\frac{\partial \Phi}{\partial z} \frac{dz}{ds} + \frac{\partial \Phi}{\partial r} \frac{dr}{ds} = \frac{d\Phi}{ds} = 0$$

from which we conclude that Φ is *constant* along the boundary

In the case of a solid shaft the integral in (11.45) is evaluated at any cross-section of radius a to give

$$T = 2\pi \int_0^a \frac{\partial \Phi}{\partial r} \, dr = 2\pi [\Phi]_0^a \tag{11.49}$$

The torque is, therefore, proportional to the difference between the values of the stress function at the boundary and at the axis of the shaft.

11.3.2 Finite element formulation

We employ triangular elements with nodes i, j and k to represent a radial section of the shaft (see Fig. 11.15). It is convenient to treat the values of $\psi = vr$, rather than v itself, as the unknowns at the nodal points. The application of the principle of virtual work (see section 6.5) to the problem therefore requires some modification.

Within a typical element e the function ψ is approximated in terms of its nodal point values $\delta^e] = [\psi_i \quad \psi_j \quad \psi_k]$ by

$$\psi = [N][\delta^e] \tag{11.50}$$

where the linear shape functions $[N]$ are defined in (9.18) with x and y replaced by r and θ. The displacement function $[u]$ can be written as

$$[u] = \mathbf{v} = r\psi = r[N][\delta^e] \tag{11.51}$$

and the element strains and stresses are

$$[e] = \begin{bmatrix} \gamma_{r\theta} \\ \gamma_{\theta z} \end{bmatrix} = r \begin{bmatrix} \dfrac{\partial \psi}{\partial r} \\[2mm] \dfrac{\partial \psi}{\partial z} \end{bmatrix} = r[B][\delta^e] \tag{11.52}$$

where

$$[B] = \frac{1}{2A} \begin{bmatrix} b_i & b_j & b_k \\ c_i & c_j & c_k \end{bmatrix} \tag{11.53}$$

and

$$[\sigma] = G[e] \tag{11.54}$$

Substituting for $[u]$ and $[e]$ from (11.51) and (11.52) into the virtual work equation (6.9), where $dS = 2\pi r \, dr$ and $dV = 2\pi r \, dr \, dz$, we obtain

$$2\pi[\delta^e]^{\mathrm{T}} \int_{S_e} r^2 [N]^{\mathrm{T}}[t] \, dr = 2\pi G[\delta^e]^{\mathrm{T}} \left[\iint_A r^3 [B]^{\mathrm{T}}[B] \, dr \, dz \right][\delta^e]$$

which leads to

$$[F^e] = [K^e][\delta^e]$$

where the element stiffness matrix is

$$[K^e] = G \int_A r^3 [B]^{\mathrm{T}}[B] \, dr \, dz \tag{11.55}$$

and the nodal point forces are

$$[F^e] = \int_{S_e} r^2 [N]^T [t] \, dr \tag{11.56}$$

After substituting for $[B]$ from (11.53) into (11.55) and integrating, we obtain

$$[K^e] = \frac{I_e G}{4A^2} \begin{bmatrix} b_i^2 + c_i^2 & b_i b_j & c_i c_j \\ b_i b_j & b_j^2 + c_j^2 & b_j b_k \\ c_i c_j & b_j b_k & b_k^2 + c_k^2 \end{bmatrix} \tag{11.57}$$

where the integral I_e is given by

$$I_e = \int_A r^3 \, dr \, dz$$
$$= \tfrac{1}{10} A (r_i^3 + r_j^3 + r_k^3 + r_i r_j r_k + r_i^2 r_j + r_j^2 r_k + r_k^2 r_i + r_i^2 r_k + r_j^2 r_i + r_k^2 r_j)$$

The nodal point forces $[F^e]$ defined in (11.56) for an element having one edge mn of length l lying in the end face $z = L$ (see Fig. 11.15) are

$$[F^e_{mn}] = - \int_{r_n}^{r_m} r^2 [N_{mn}]^T \tau_{\theta z} \, dr \tag{11.58}$$

where

$$[N_{mn}] = [N_m \quad N_n] = \begin{bmatrix} 1 - \dfrac{r}{l} & \dfrac{r}{l} \end{bmatrix}$$

If the prescribed shear stress $\tau_{\theta z}$ on the end face varies *linearly* along mn, we can write

$$\tau_{\theta z} = [N_{mn}][\bar{\tau}_{\theta z}] \tag{11.59}$$

where $[\bar{\tau}_{\theta z}] = [(\tau_{\theta z})_m \ (\tau_{\theta z})_n]^T$ are the values of $\tau_{\theta z}$ at the nodes m and n. Substituting (11.59) into (11.58) and integrating, we obtain

$$[F^e_{mn}] = - \frac{l}{60} \begin{bmatrix} 12r_m^2 + 6r_m r_n + 2r_n^2 & 3r_m^2 + 4r_m r_n + 3r_n^2 \\ 3r_m^2 + 4r_m r_n + 3r_n^2 & 2r_m^2 + 6r_m r_n + 12r_n^2 \end{bmatrix} \begin{bmatrix} (\tau_{\theta z})_m \\ (\tau_{\theta z})_n \end{bmatrix} \tag{11.60}$$

The element stresses are obtained by substituting (11.50) into (11.40) to give

$$\begin{aligned}
\tau_{r\theta} &= Gr \frac{\partial \psi}{\partial r} = \frac{Gr}{2A} (b_i \psi_i + b_j \psi_j + b_k \psi_k) \\
\tau_{\theta z} &= Gr \frac{\partial \psi}{\partial z} = \frac{Gr}{2A} (c_i \psi_i + c_j \psi_j + c_k \psi_k)
\end{aligned} \tag{11.61}$$

11.3.3 Use of FIESTA2

To select the appropriate element subroutine ELEM6 in FIESTA2 we put IETYPE=6 in the data file (see Fig. 7.2). The automatic mesh generation facility (see section 7.3) is available if required. The array XY stores the nodal point coordinates in the rz plane.

Subroutine ELEM6 calculates the element stiffness array SE using (11.57), and the array FE of nodal point forces corresponding to a linearly varying shear stress $\tau_{\theta z}$ on a plane end face of the shaft $z = L$ using (11.60). The centroidal values of the element stresses are calculated in subroutine STR6 using (11.61), with $r = r_c$, and are stored in the array SS.

The material data for the problem are specified in the manner described in section 7.2.2. With NOPROP=1 only a single property EDAT(ISET,1)=G is to be specified for each property set ISET.

Stress boundary conditions are specified in the manner described in section 7.2.3. The prescribed values for stress boundary condition IBC are stored in the array PS as follows:

$$PS(IBC,1) = (\tau_{\theta z})_m, \qquad PS(IBC,2) = 0$$
$$PS(IBC,3) = (\tau_{\theta z})_n, \qquad PS(IBC,4) = 0$$

Sufficient displacement boundary conditions must be prescribed to ensure that $[K]$ is a nonsingular matrix.

11.3.4 Stepped shaft with shoulder fillets

Where a circular shaft undergoes an abrupt change in diameter, a shoulder with rounded fillets of radius R is used to lessen the resulting stress concentration (see Fig. 11.16). For $a = 0.9b$ and $R = 0.1b$ the problem is modelled using the finite element mesh shown in Fig. 11.17. A torque T is applied by fixing the end $z = 0$ and prescribing a suitable distribution of the shear stress $\tau_{\theta z}$ at $z = L$. Assuming that stresses in the shaft near this end are given by the elementary theory of torsion, it follows from equations (11.1) that we must prescribe

$$\tau_{\theta z} = \frac{Tr}{J} = \frac{2Tr}{\pi b^3} \tag{11.62}$$

along a radius at $z = L$.

The variation in τ_{max}/τ_0 in the elements bordering the surface of the shaft is shown in Fig. 11.18, where $\tau_0 = 2T/\pi a^3$. The stress distribution is only disturbed

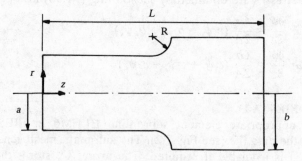

Fig. 11.16—Stepped shaft with shoulder fillets.

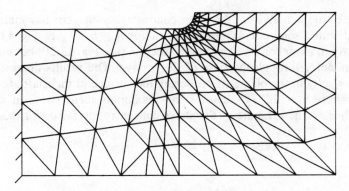

Fig. 11.17—Finite element mesh for a stepped shaft.

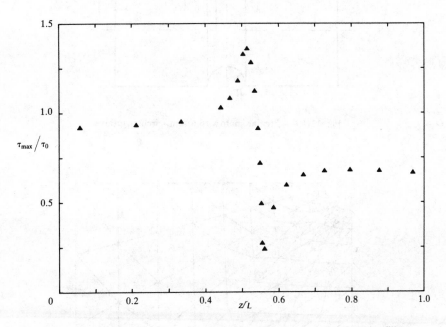

Fig. 11.18—Variation in the maximum shear stress τ_{max} around a shoulder fillet.

locally by the fillet and at a sufficient distance from the shoulder returns to that for a shaft of uniform cross-section. The stress concentration factor, corresponding to the maximum value of the ratio τ_{max}/τ_0, is equal to 1.36.

11.3.5 Shaft with a semi-circular groove
Grooves are a common feature in a range of design configurations. Examples include the use of a groove to provide stress relief at a shoulder where it is not possible to use a fillet of suitable radius, and a groove to take a snap ring used to locate a bearing.

In a finite element analysis of the stress concentration in a circular shaft of radius a associated with a semi-circular groove of radius $R = b/10$ (see Fig. 11.19), it is only necessary to model one-half of a radial section (see Fig. 11.20) bounded by the plane of symmetry through the base of the groove. The results obtained for the variation in τ_{max}/τ_0 in the elements bordering the surface of the shaft are shown in Fig. 11.21 where $\tau_0 = 2T/\pi a^3$ denotes the maximum shear stress in a shaft of uniform radius a. The estimate of the stress concentration factor for this problem is 1.61.

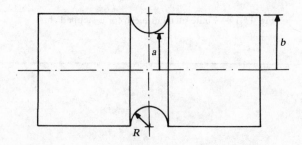

Fig. 11.19—Circular shaft with a semi-circular groove.

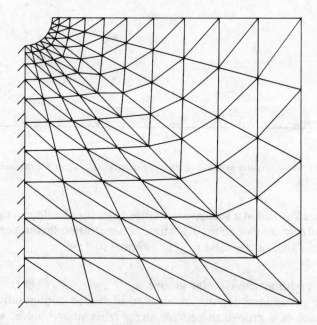

Fig. 11.20—Finite element mesh for circular shaft with a semi-circular groove.

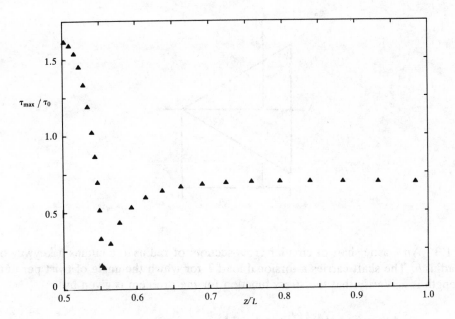

Fig. 11.21—Variation in the maximum shear stress τ_{max} around a groove.

PROBLEMS

11.1 Show that, for the same angle of twist, an elastic shaft of elliptical cross-section carries a greater maximum shear stress than a shaft of circular cross-section whose diameter is equal to the length of the minor axis of the ellipse. Which shaft carries the greater torque for the same maximum shear stress?

(Answer: the shaft of elliptical cross-section.)

11.2 An elastic shaft with a cross-section in the form of an equilateral triangle of height h carries a torsional load. Verify that the stress function for the problem is given by

$$\Phi = m\left[(x^2 + y^2) - \frac{1}{h}(x^3 - 3xy^2) - \frac{4h^2}{27}\right]$$

where the constant m has the value $-\alpha G/2$. Obtain expressions for the torsional constant $T/\alpha G$ and the maximum shear stress at the point A.

(Answers: $h^4/15\sqrt{3}$, $\alpha Gh/2$.)

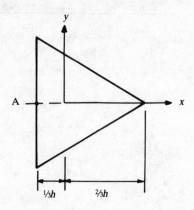

11.3   An elastic shaft of circular cross-section, of radius a, contains a keyway of radius b. The shaft carries a torsional load T for which the angle of twist per unit length is α. Verify that the stress function for the problem is given by

$$\Phi = m(r^2 - b^2)\left(\frac{2a}{r}\cos\theta - 1\right)$$

where the constant m has the value $\alpha G/2$, and obtain an expression for the maximum shear stress at the point A.

(Answer: $\alpha G(2a - b)$.)

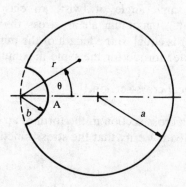

11.4   For an elastic shaft of rectangular cross-section, breadth b and depth d, where $b \gg d$, the stress function can be assumed to be independent of x. Find the solution for Φ and the torsional constant $T/\alpha G$.

(Answers: $\alpha G(d^2 - 4y^2)/4, bd^3/3$.)

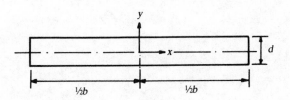

11.5 An elastic uniformly tapering shaft of circular cross-section transmits a torque T. Show that the stress function for the problem is given by

$$\Phi = m\left(\frac{z}{R} + \frac{\lambda z^3}{R^3}\right)$$

where $R = \sqrt{r^2 + z^2}$, and $\lambda = -\frac{1}{3}$, $m = -3T/2\pi(\cos^3\alpha - 3\cos\alpha + 2)$.

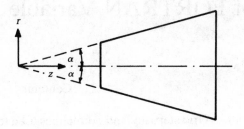

APPENDIX A1
Glossary of FORTRAN Variable Names

Variable	Contents
DR, DS	Grid steps in r and s directions used to generate a uniform square mesh.
DS	Array storing diagonal coefficients of $[K]$.
EDAT(IPROP, ISET)	Material property IPROP for data set ISET.
FE(IDE)	Prescribed force for element degree of freedom IDE.
FS(IDS)	Prescribed force for system degree of freedom IDS.
I	Forward reduction step number.
IBC	Boundary condition number.
IBW	Semi-bandwidth of system stiffness matrix.
ICO	Coordinate number (=1, 2, . . ., NOCO).
IDA(NS,IDN)	Active equation number corresponding to degree of freedom IDN of node NS.
IDE	Element degree of freedom (=1, 2, . . ., NODE).
IDN	Nodal degree of freedom (=1, 2, . . ., NODN).
IDS	System degree of freedom (=1, 2,. . ., NODS).
IE	Element number (=1, 2, . . ., NOE).
IEQ	Active equation number (=1, 2, . . ., NOEQ).
IETYPE	Element type.
IE1, IE2, IE3	Element numbers in a repeating unit used in automatic mesh generation.

IFLAG(IR)	Flag for generation of element row IR: 0, no change of element size; 1 halve element size; 2, double element size.
IJ(IE, NE)	Nodal point NE for element IE.
IPAR	Array of element parameters.
IPD	Pointers for displacement boundary condition IBC: IPD(IBC, 1), node; IPD(IBC, 2), degree of freedom.
IPF	Pointers for force boundary condition IBC: IPC(IBC, 1), node; IPF(IBC, 2), degree of freedom.
IPROP	Element property.
IPS	Pointers for stress boundary condition IBC: IPS(IBC, 1), element number; IPS(IBC, 2), IPS(IBC, 3), nodes defining element edge.
IR	Element row.
IR, IC	Row, column positions in matrix $[K]$.
IRC(IDE)	Row or column position in $[K]$ equivalent to row or column position IDE in $[K^e]$.
IRE, ICE	Row, column position in element stiffness array.
IRS, ICS	Row, column position in system stiffness array.
ISET	Element data set number ($=1, 2, \ldots$, NOSET).
IST	Element stress component number ($=1, 2, \ldots$, NOST).
ISUB	Subregion number ($=1, 2, \ldots$, NOSUB).
IUNIT	Repeating unit number in mesh generation.
JPS(IE)	Stress boundary condition number for element IE.
MAT(IE)	Material set number for element IE.
MFLAG	Mesh data flag: 0, read mesh data from file; 1, automatic generation of mesh data.
NE	Element node ($=1, 2, \ldots$, NONE).
NOCO	Number of coordinates defining element geometry.
NODE	Number of element degrees of freedom.
NODN	Number of nodal degrees of freedom.
NODS	Number of system degrees of freedom.
NOE	Number of elements in the system.
NOEQ	Number of active equations.
NOES	Number of rows of elements in the s direction.
NONE	Number of nodes per element.
NONR	Number of nodes in the r direction.
NONS	Number of nodes in the system.
NOPD	Number of prescribed displacements.
NOPF	Number of prescribed forces.
NOPROP	Number of properties per set of material data.

NOPS	Number of prescribed stresses.
NOSET	Number of sets of material data.
NOST	Number of element stress components.
NOSUB	Number of subregions used to generate mesh.
NS	System node (=1, 2, . . ., NONS).
NSTART	First node of current subregion.
N1, N2	Nodes for a repeating unit in mesh generation.
PD(IBC)	Prescribed displacement for boundary condition IBC.
PF(IBC)	Prescribed force for boundary condition IBC.
PS	Prescribed stresses for boundary condition IBC:
	PS(IBC, 1), PS(IBC, 2), normal shear stress at
	node m; PS(IBC, 3), PS(IBC, 4), normal shear
	stress at node n.
R	Elimination ratio for forward reduction of $[K]$.
R, S	Nodal coordinates for uniform square mesh.
SE(IRE,ICE)	Stiffness coefficient in row IRE and column ICE of
	element stiffness array.
SF	Shape functions defining coordinate mappings used in
	mesh generation.
SS	System stiffness coefficients, and element stresses:
	SS(IRS, ICS), system stiffness coefficient;
	SS(IE, IST), stress component IST for element IE.
	IE.
TITLE	Problem title.
XY(NS, ICO)	Coordinate ICO for nodal point NS.
XYSUB	Coordinates of points defining the boundaries of
	subregions used in mesh generation.

Appendix A2

Listing of FIESTA1

```
      PROGRAM FIESTA1
C     *****      FINITE ELEMENT STRESS ANALYSIS       *****
C     *****   OF ONE-DIMENSIONAL ELASTIC PROBLEM      *****
C     *****                SEE                        *****
C     *****        "ENGINEERING STRESS ANALYSIS"      *****
C     *****   BY D.N.FENNER (ELLIS HORWOOD 1987)      *****
      CALL SPEC
      CALL LIST
      CALL ASSEM
      CALL SOLVE
      CALL STRESS
      CALL OUTPUT
      END

      SUBROUTINE SPEC
C     *****   READ PROBLEM SPECIFICATION : SEE SECTION 5.3.1   *****
      COMMON/GEOM/D0,D1,FL
      COMMON/ELEM/NOE,IJ(100,2),NODN,NONE,NODE
      COMMON/NODE/NONS,XY(100,1),NOCO
      COMMON/MATS/E
      COMMON/EQNS/NODS,NOEQ,IBW,SS(100,2),FS(100)
      COMMON/DBCS/NOPD,IPD(5,2),PD(5)
      COMMON/FBCS/NOPF,IPF(5,2),PF(5)
C     GEOMETRY AND MATERIAL
      READ(5,*)D0,D1,FL,E
C     ELEMENT PARAMETERS
      READ(5,*)NONE,NODN,NOCO
      NODE=NONE*NODN
C     DISCRETISATION
      READ(5,*)NOE,NONS
      NODS=NONS*NODN
      DO 10 NS=1,NONS
   10 XY(NS,1)=FL*FLOAT(NS-1)/FLOAT(NOE)
      DO 20 IE=1,NOE
      IJ(IE,1)=IE
   20 IJ(IE,2)=IE+1
C     PRESCRIBED FORCES
      READ(5,*)NOPF
      DO 30 IBC=1,NOPF
   30 READ(5,*)IPF(IBC,1),IPF(IBC,2),PF(IBC)
C     PRESCRIBED DISPLACEMENTS
      READ(5,*)NOPD
      DO 40 IBC=1,NOPD
   40 READ(5,*)IPD(IBC,1),IPD(IBC,2),PD(IBC)
      NOEQ=NODS
      IBW=2
      RETURN
      END

      SUBROUTINE LIST
C     *****  LIST PROBLEM SPECIFICATION  *****
      COMMON/MATS/E
      COMMON/GEOM/D0,D1,FL
      COMMON/ELEM/NOE,IJ(100,2),NODN,NONE,NODE
      COMMON/NODE/NONS,XY(100,1),NOCO
      COMMON/EQNS/NODS,NOEQ,IBW,SS(100,2),FS(100)
      COMMON/DBCS/NOPD,IPD(5,2),PD(5)
      COMMON/FBCS/NOPF,IPF(5,2),PF(5)
      WRITE(6,600)
      WRITE(6,610)D0,D1,FL,E
      WRITE(6,620)NOE,NONS
      WRITE(6,630)
```

```
      WRITE(6,640)(IPF(IBC,1),PF(IBC),IBC=1,NOPF)
      WRITE(6,650)
      WRITE(6,640)(IPD(IBC,1),PD(IBC),IBC=1,NOPD)
  600 FORMAT(' FIESTA1 : VERSION 1.0'/
     * ' SEE "ENGINEERING STRESS ANALYSIS" BY D.N.FENNER'/
     * ' ELLIS HORWOOD (1987)'///
     * ' PROBLEM SPECIFICATION'//)
  610 FORMAT('     BAR DIAMETER D0   = ',E12.4/
     *       '     BAR DIAMETER D1   = ',E12.4/
     *       '     BAR LENGTH L      = ',E12.4/
     *       '     ELASTIC MODULUS E = ',E12.4//)
  620 FORMAT('     NUMBER OF ELEMENTS = ',I5/
     *       '     NUMBER OF NODES    = ',I5//)
  630 FORMAT(' FORCE BOUNDARY CONDITIONS'//
     *       '     NODE          FORCE'/)
  640 FORMAT(I5,E15.4)
  650 FORMAT(//' DISPLACEMENT BOUNDARY CONDITIONS'//
     *       '     NODE          DISP'/)
      RETURN
      END


      SUBROUTINE ASSEM
C ***** ASSEMBLE EQUILIBRIUM EQUATIONS : SEE SECTION 5.3.3 *****
      COMMON/ELEM/NOE,IJ(100,2),NODN,NONE,NODE
      COMMON/EQNS/NODS,NOEQ,IBW,SS(100,2),FS(100)
      COMMON/FBCS/NOPF,IPF(5,2),PF(5)
      COMMON/DBCS/NOPD,IPD(5,2),PD(5)
      DIMENSION SE(2,2),IRC(2)
C ZERO ARRAYS
      DO 10 IRS=1,NOEQ
      FS(IRS)=0.
      DO 10 ICS=1,IBW
   10 SS(IRS,ICS)=0.
C ASSEMBLE SYSTEM STIFFNESS ARRAY
      DO 50 IE=1,NOE
      CALL ELEMNT(IE,SE)
      DO 20 NE=1,NONE
   20 IRC(NE)=IJ(IE,NE)
      DO 40 IRE=1,NODE
      IRS=IRC(IRE)
      DO 30 ICE=1,NODE
      ICS=IRC(ICE)-IRS+1
      IF(ICS.LT.1) GO TO 30
      SS(IRS,ICS)=SS(IRS,ICS)+SE(IRE,ICE)
   30 CONTINUE
   40 CONTINUE
   50 CONTINUE
C ASSEMBLE NODAL POINT FORCES ARRAY
      DO 60 IBC=1,NOPF
      IDS=IPF(IBC,1)
   60 FS(IDS)=FS(IDS)+PF(IBC)
C DISPLACEMENT BOUNDARY CONDITIONS : SEE SECTION 5.3.4
      DO 70 IBC=1,NOPD
      IDS=IPD(IBC,1)
      SS(IDS,1)=SS(IDS,1)*1.E+20
   70 FS(IDS)=PD(IBC)*SS(IDS,1)
      RETURN
      END


      SUBROUTINE ELEMNT(IE,SE)
C ***** STIFFNESS MATRIX FOR BAR ELEMENT  *****
C *****          SEE SECTION 5.3.2         *****
      COMMON/GEOM/D0,D1,FL
```

```
      COMMON/ELEM/NOE,IJ(100,2),NODN,NONE,NODE
      COMMON/NODE/NONS,XY(100,1),NOCO
      COMMON/MATS/E
      DIMENSION SE(2,2)
      I=IJ(IE,1)
      J=IJ(IE,2)
      FLE=XY(J,1)-XY(I,1)
      XMEAN=(XY(I,1)+XY(J,1))/2.
      DMEAN=D0+XMEAN/FL*(D1-D0)
      A=4.*ATAN(1.)*DMEAN**2/4.
      S=E*A/FLE
      SE(1,1)= S
      SE(1,2)=-S
      SE(2,1)=-S
      SE(2,2)= S
      RETURN
      END

      SUBROUTINE SOLVE
C  *****   SOLVE SET OF TRIDIAGONAL SYMMETRIC EQUATIONS    *****
C  *****   USING GAUSSIAN ELIMINATION : SEE SECTION 5.3.5  *****
      COMMON/EQNS/NODS,NOEQ,IBW,SS(100,2),FS(100)
      DIMENSION DS(100)
      DO 10 IEQ=1,NOEQ
   10 DS(IEQ)=SS(IEQ,1)
C  FORWARD REDUCTION : SEE (5.22)
      DO 20 I=1,NOEQ-1
      R=SS(I,2)/SS(I,1)
      SS(I+1,1)=SS(I+1,1)-R*SS(I,2)
      IF(SS(I+1,1)/DS(I+1).LT.1E-3) GO TO 40
   20 FS(I+1)=FS(I+1)-R*FS(I)
C  BACK SUBSTITUTION : SEE (5.24),(5.25)
      FS(NOEQ)=FS(NOEQ)/SS(NOEQ,1)
      DO 30 I=NOEQ-1,1,-1
   30 FS(I)=(FS(I)-SS(I,2)*FS(I+1))/SS(I,1)
      RETURN
   40 WRITE(6,600)
  600 FORMAT(//'    ***** ILL-CONDITIONED EQUATIONS *****')
      STOP
      END

      SUBROUTINE STRESS
C  *****   CALCULATE ELEMENT STRESSES : SEE SECTION 5.3.6  *****
      COMMON/ELEM/NOE,IJ(100,2),NODN,NONE,NODE
      COMMON/NODE/NONS,XY(100,1),NOCO
      COMMON/MATS/E
      COMMON/EQNS/NODS,NOEQ,IBW,SS(100,2),FS(100)
      DO 10 IE=1,NOE
      I=IJ(IE,1)
      J=IJ(IE,2)
   10 SS(IE,1)=E*(FS(J)-FS(I))/(XY(J,1)-XY(I,1))
      RETURN
      END

      SUBROUTINE OUTPUT
C  *****   LIST PROBLEM SOLUTION   *****
      COMMON/ELEM/NOE,IJ(100,2),NODN,NONE,NODE
      COMMON/NODE/NONS,XY(100,1),NOCO
      COMMON/EQNS/NODS,NOEQ,IBW,SS(100,2),FS(100)
      WRITE(6,600)
      WRITE(6,610)
      WRITE(6,620) (NS,FS(NS),NS=1,NONS)
      WRITE(6,630)
      WRITE(6,620) (IE,SS(IE,1),IE=1,NOE)
```

```
600 FORMAT(///' PROBLEM SOLUTION'//)
610 FORMAT(' NODAL POINT DISPLACEMENTS'//
   *        '   NODE       DISP'/)
620 FORMAT(I5,E15.4)
630 FORMAT(///' ELEMENT STRESSES'//
   *        '   ELEM      STRESS'/)

RETURN
END
```

Appendix A3
Listing of FIESTA2

```
      PROGRAM FIESTA2
C     *****      FINITE ELEMENT STRESS ANALYSIS      *****
C     *****   OF TWO-DIMENSIONAL ELASTIC PROBLEMS    *****
C     *****               SEE                        *****
C     *****        "ENGINEERING STRESS ANALYSIS"     *****
C     *****   BY D.N.FENNER (ELLIS HORWOOD 1987)     *****
      CALL SPEC
      CALL ACTIVE
      CALL BWIDTH
      CALL LIST
      CALL ASSEM
      CALL SOLVE
      CALL STRESS
      CALL OUTPUT
      END

      SUBROUTINE SPEC
C     *****   PROBLEM SPECIFICATION : SEE SECTION 7.2   *****
      CHARACTER*80 TITLE
      DIMENSION IPAR(6,5)
      COMMON/ELEM/IETYPE,NOE,IJ(1000,3),MAT(1000),NODN,NONE,NODE,NOST
      COMMON/NODE/NONS,XY(600,2),NOCO
      COMMON/EQNS/NODS,NOEQ,IBW,IDA(600,6),SS(1200,50),FS(1200)
      COMMON/DBCS/NOPD,IPD(50,2),PD(50)
      COMMON/NAME/TITLE
      COMMON/MATS/NOPROP,NOSET,EDAT(5,5)
      DATA (IPAR(1,I),I=1,5)/2,2,2,1,2/
      DATA (IPAR(2,I),I=1,5)/3,2,2,6,3/
      DATA (IPAR(3,I),I=1,5)/2,3,2,3,4/
      DATA (IPAR(4,I),I=1,5)/2,3,2,4,3/
      DATA (IPAR(5,I),I=1,5)/1,3,2,2,0/
      DATA (IPAR(6,I),I=1,5)/1,3,2,2,1/
      READ(5,*)TITLE
      READ(5,*)IETYPE
      NODN=IPAR(IETYPE,1)
      NONE=IPAR(IETYPE,2)
      NOCO=IPAR(IETYPE,3)
      NOST=IPAR(IETYPE,4)
      NOPROP=IPAR(IETYPE,5)
      NODE=NODN*NONE
      READ(5,*)MFLAG
C     MESH DATA
      IF(MFLAG.EQ.0) CALL RMESH
      IF(MFLAG.EQ.1) CALL GMESH
      NODS=NONS*NODN
C     MATERIAL DATA
      CALL PROPS
C     BOUNDARY CONDITIONS
      CALL BOUND
      RETURN
      END

      SUBROUTINE RMESH
C     *****   READ MESH SPECIFICATION : SEE SECTION 7.2.1   *****
      COMMON/ELEM/IETYPE,NOE,IJ(1000,3),MAT(1000),NODN,NONE,NODE,NOST
      COMMON/NODE/NONS,XY(600,2),NOCO
      READ(5,*)NONS
      DO 10 NS=1,NONS
   10 READ(5,*) (XY(NS,ICO),ICO=1,NOCO)
      READ(5,*)NOE
      DO 20 IE=1,NOE
   20 READ(5,*) (IJ(IE,NE),NE=1,NONE)
      RETURN
      END
```

```
      SUBROUTINE GMESH
C  *****  GENERATE MESH SPECIFICATION  : SEE SECTION 7.3  *****
      COMMON/ELEM/IETYPE,NOE,IJ(1000,3),MAT(1000),NODN,NONE,NODE,NOST
      COMMON/NODE/NONS,XY(600,2),NOCO
      NOE=0
      NONS=0
      READ(5,*)NOSUB
      DO 10 ISUB=1,NOSUB
      CALL SQUARE(NSTART)
   10 CALL MAP(NSTART)
      RETURN
      END


      SUBROUTINE SQUARE(NSTART)
C  *****  GENERATE SQUARE MESH : SEE SECTION 7.3.1  *****
      COMMON/NODE/NONS,XY(600,2),NOCO
      DIMENSION IFLAG(50)
      READ(5,*)NONR,NOES
      READ(5,*)(IFLAG(IR),IR=1,NOES)
      NSTART=NONS+1
      DR=2./REAL(NONR-1)
      DS=2./REAL(NOES)
      S=-1.
      IF(NONS.EQ.0) CALL NLINE(NONR,DR,S)
      DO 10 IR=1,NOES
      IF(IFLAG(IR).EQ.0) CALL EROW(NONR,S,DS)
      IF(IFLAG(IR).EQ.1) CALL EROW1(NONR,DR,S,DS)
      IF(IFLAG(IR).EQ.2) CALL EROW2(NONR,DR,S,DS)
   10 CALL NLINE(NONR,DR,S)
      SMAX=XY(NONS,2)
      IF(SMAX.EQ.1.) RETURN
      DO 20 NS=NSTART,NONS
   20 XY(NS,2)=(2.*XY(NS,2)+1.-SMAX)/(1.+SMAX)
      RETURN
      END


      SUBROUTINE NLINE(NONR,DR,S)
C  *****  GENERATE COORDINATES OF A LINE OF NODES  *****
      COMMON/NODE/NONS,XY(600,2),NOCO
      DO 10 I=1,NONR
      NS=NONS+I
      XY(NS,1)=-1.+DR*REAL(I-1)
   10 XY(NS,2)=S
      NONS=NONS+NONR
      RETURN
      END


      SUBROUTINE EROW(NONR,S,DS)
C  *****  GENERATE NORMAL ROW OF ELEMENTS : SEE SECTION 7.3.1  *****
      COMMON/ELEM/IETYPE,NOE,IJ(1000,3),MAT(1000),NODN,NONE,NODE,NOST
      COMMON/NODE/NONS,XY(600,2),NOCO
      DO 10 IUNIT=1,NONR-1
      N1=NONS-NONR+IUNIT
      N2=NONS+IUNIT
      IE1=NOE+2*IUNIT-1
      IE2=IE1+1
      IJ(IE1,1)=N1
      IJ(IE1,2)=N2+1
      IJ(IE1,3)=N2
      IJ(IE2,1)=N1
      IJ(IE2,2)=N1+1
   10 IJ(IE2,3)=N2+1
      NOE=NOE+2*(NONR-1)
      S=S+DS
      RETURN
```

```
      END
      SUBROUTINE EROW1(NONR,DR,S,DS)
C *****  GENERATE TRANSITIONAL ROW OF ELEMENTS TO DOUBLE  *****
C *****        SIZE OF ELEMENTS : SEE SECTION 7.3.2        *****
      COMMON/ELEM/IETYPE,NOE,IJ(1000,3),MAT(1000),NODN,NONE,NODE,NOST
      COMMON/NODE/NONS,XY(600,2),NOCO
      DO 10 IUNIT=1,(NONR-1)/2
      N1=NONS-NONR+2*IUNIT-1
      N2=NONS+IUNIT
      IE1=NOE+3*IUNIT-2
      IE2=IE1+1
      IE3=IE2+1
      IJ(IE1,1)=N1
      IJ(IE1,2)=N1+1
      IJ(IE1,3)=N2
      IJ(IE2,1)=N1+1
      IJ(IE2,2)=N2+1
      IJ(IE2,3)=N2
      IJ(IE3,1)=N1+1
      IJ(IE3,2)=N1+2
   10 IJ(IE3,3)=N2+1
      NOE=NOE+3*(NONR-1)/2
      NONR=(NONR-1)/2+1
      S=S+1.5*DS
      DR=DR*2.
      DS=2*DS
      RETURN
      END

      SUBROUTINE EROW2(NONR,DR,S,DS)
C *****  GENERATE TRANSITIONAL ROW OF ELEMENTS TO HALVE  *****
C *****        SIZE OF ELEMENTS : SEE SECTION 7.3.2       *****
      COMMON/ELEM/IETYPE,NOE,IJ(1000,3),MAT(1000),NODN,NONE,NODE,NOST
      COMMON/NODE/NONS,XY(600,2),NOCO
      DO 10 IUNIT=1,NONR-1
      N1=NONS-NONR+IUNIT
      N2=NONS+2*IUNIT-1
      IE1=NOE+3*IUNIT-2
      IE2=IE1+1
      IE3=IE2+1
      IJ(IE1,1)=N1
      IJ(IE1,2)=N2+1
      IJ(IE1,3)=N2
      IJ(IE2,1)=N1
      IJ(IE2,2)=N1+1
      IJ(IE2,3)=N2+1
      IJ(IE3,1)=N1+1
      IJ(IE3,2)=N2+2
   10 IJ(IE3,3)=N2+1
      NOE=NOE+3*(NONR-1)
      NONR=2*NONR-1
      S=S+.75*DS
      DR=DR/2.
      DS=.5*DS
      RETURN
      END

      SUBROUTINE MAP(NSTART)
C *****  PERFORM QUADRATIC MAPPINGS : SEE SECTION 7.3.3  *****
      COMMON/NODE/NONS,XY(600,2),NOCO
      DIMENSION SF(8),XYSUB(8,2)
      READ(5,*)((XYSUB(I,J),J=1,2),I=1,8)
      DO 20 NS=NSTART,NONS
      R=XY(NS,1)
```

H

```
        S=XY(NS,2)
        CALL SHAPE(R,S,SF)
        XY(NS,1)=0.
        XY(NS,2)=0.
        DO 10 I=1,8
        XY(NS,1)=XY(NS,1)+SF(I)*XYSUB(I,1)
     10 XY(NS,2)=XY(NS,2)+SF(I)*XYSUB(I,2)
     20 CONTINUE
        RETURN
        END


        SUBROUTINE SHAPE(R,S,SF)
C  *****   SHAPE FUNCTIONS FOR QUADRATIC MAPPINGS : SEE (7.5)   *****
        DIMENSION SF(8)
        SF(1)=.25*(1+R)*(1+S)*(R+S-1)
        SF(2)=.5*(1-R*R)*(1+S)
        SF(3)=.25*(1-R)*(1+S)*(-R+S-1)
        SF(4)=.5*(1-R)*(1-S*S)
        SF(5)=.25*(1-R)*(1-S)*(-R-S-1)
        SF(6)=.5*(1-R*R)*(1-S)
        SF(7)=.25*(1+R)*(1-S)*(R-S-1)
        SF(8)=.5*(1+R)*(1-S*S)
        RETURN
        END


        SUBROUTINE PROPS
C  *****   READ MATERIAL PROPERTIES : SEE SECTION 7.2.2   *****
        COMMON/ELEM/IETYPE,NOE,IJ(1000,3),MAT(1000),NODN,NONE,NODE,NOST
        COMMON/MATS/NOPROP,NOSET,EDAT(5,5)
        DO 10 IE=1,NOE
     10 MAT(IE)=0
        READ(5,*)NOSET
        IF(NOSET.EQ.0) RETURN
        READ(5,*)((EDAT(ISET,IPROP),IPROP=1,NOPROP),ISET=1,NOSET)
        IF(NOSET.EQ.1) THEN
        DO 20 IE=1,NOE
     20 MAT(IE)=1
        ELSE
        READ(5,*)(MAT(IE),IE=1,NOE)
        ENDIF
        RETURN
        END


        SUBROUTINE BOUND
C  *****   READ BOUNDARY CONDITIONS : SEE SECTION 7.2.3   *****
        COMMON/ELEM/IETYPE,NOE,IJ(1000,3),MAT(1000),NODN,NONE,NODE,NOST
        COMMON/DBCS/NOPD,IPD(50,2),PD(50)
        COMMON/SBCS/NOPS,IPS(50,3),PS(50,4),JPS(1000)
        COMMON/FBCS/NOPF,IPF(20,2),PF(20)
C  DISPLACEMENTS
        READ(5,*)NOPD
        IF(NOPD.NE.0)READ(5,*)((IPD(IBC,J),J=1,2),PD(IBC),IBC=1,NOPD)
        DO 10 IE=1,NOE
     10 JPS(IE)=0
C  STRESSES
        READ(5,*)NOPS
        IF(NOPS.NE.0) THEN
        DO 20 IBC=1,NOPS
        READ(5,*)(IPS(IBC,J),J=1,3),(PS(IBC,J),J=1,4)
        IE=IPS(IBC,1)
     20 JPS(IE)=IBC
        ENDIF
C  FORCES
        READ(5,*)NOPF
        IF(NOPF.NE.0) THEN
```

```
      READ(5,*)((IPF(IBC,J),J=1,2),PF(IBC),IBC=1,NOPF)
      ENDIF
      RETURN
      END

      SUBROUTINE LIST
C   *****  LIST PROBLEM SPECIFICATION  *****
      COMMON/DBCS/NOPD,IPD(50,2),PD(50)
      COMMON/FBCS/NOPF,IPF(20,2),PF(20)
      COMMON/SBCS/NOPS,IPS(50,3),PS(50,4),JPS(1000)
      COMMON/ELEM/IETYPE,NOE,IJ(1000,3),MAT(1000),NODN,NONE,NODE,NOST
      COMMON/NODE/NONS,XY(600,2),NOCO
      COMMON/EQNS/NODS,NOEQ,IBW,IDA(600,6),SS(1200,50),FS(1200)
      COMMON/MATS/NOPROP,NOSET,EDAT(5,5)
      COMMON/NAME/TITLE
      CHARACTER*80 TITLE
      WRITE(6,600)TITLE
C   PROBLEM PARAMETERS
      WRITE(6,610)IETYPE,NOE,NONS,NODS,NOEQ,IBW
C   NODAL POINT DATA
      WRITE(6,620)
      IF(IETYPE.EQ.4.OR.IETYPE.EQ.6) THEN
      WRITE(6,630)
      ELSE
      WRITE(6,640)
      ENDIF
      WRITE(6,650)(NS,(XY(NS,ICO),ICO=1,2),NS=1,NONS)
C   ELEMENT DATA
      WRITE(6,660)
      IF(NONE.EQ.2) THEN
      WRITE(6,670)
      WRITE(6,680)(IE,(IJ(IE,NE),NE=1,2),MAT(IE),JPS(IE),IE=1,NOE)
      ENDIF
      IF(NONE.EQ.3) THEN
      WRITE(6,690)
      WRITE(6,700)(IE,(IJ(IE,NE),NE=1,3),MAT(IE),JPS(IE),IE=1,NOE)
      ENDIF
      IF(IETYPE.EQ.5) RETURN
      IF(IETYPE.EQ.1) WRITE(6,710)
      IF(IETYPE.EQ.2) WRITE(6,720)
      IF(IETYPE.EQ.3) WRITE(6,730)
      IF(IETYPE.EQ.4) WRITE(6,740)
      IF(IETYPE.EQ.6) WRITE(6,750)
      DO 10 ISET=1,NOSET
   10 WRITE(6,760) ISET,(EDAT(ISET,IPROP),IPROP=1,NOPROP)
C   DISPLACEMENT BOUNDARY CONDITIONS
      IF(NOPD.GT.0) THEN
      WRITE(6,770)
      WRITE(6,780)
      WRITE(6,790)(IBC,IPD(IBC,1),IPD(IBC,2),PD(IBC),IBC=1,NOPD)
      ENDIF
C   FORCE BOUNDARY CONDITIONS
      IF(NOPF.GT.0) THEN
      WRITE(6,800)
      WRITE(6,780)
      WRITE(6,790)(IBC,IPF(IBC,1),IPF(IBC,2),PF(IBC),IBC=1,NOPF)
      ENDIF
C   STRESS BOUNDARY CONDITIONS
      IF(NOPS.GT.0) THEN
      WRITE(6,810)
      WRITE(6,820)
      WRITE(6,830)(IBC,(IPS(IBC,I),I=1,3),(PS(IBC,I),I=1,4),IBC=1,NOPS)
      ENDIF
  600 FORMAT(' FIESTA2 : VERSION 1.0'/
     * ' SEE "ENGINEERING STRESS ANALYSIS" BY D.N.FENNER'/
```

```
     * ' ELLIS HORWOOD (1987)'////A80///
     * ' PROBLEM SPECIFICATION'/' *********************'//)
 610 FORMAT(' ELEMENT TYPE                = ',I5/
     *        ' NO. OF ELEMENTS            = ',I5/
     *        ' NO. OF NODAL POINTS        = ',I5/
     *        ' TOTAL NO. OF EQUATIONS     = ',I5/
     *        ' NO. OF ACTIVE EQUATIONS    = ',I5/
     *        ' SEMI-BANDWIDTH             = ',I5)
 620 FORMAT(///' NODAL POINT DATA'/)
 630 FORMAT( 3(' NODE',5X,'R',10X,'Z',4X)/)
 640 FORMAT( 3(' NODE',5X,'X',10X,'Y',4X)/)
 650 FORMAT(3(1X,I3,2E11.3))
 660 FORMAT(///' ELEMENT DATA'/)
 670 FORMAT(3('  ELEM  I   J  MAT BC')/)
 680 FORMAT(3(1X,3I4,1X,2I3,1X))
 690 FORMAT(3('  ELEM  I   J   K  MAT BC')/)
 700 FORMAT(3(1X,4I4,1X,2I3,1X))
 710 FORMAT(//'   MAT',6X,'E',11X,'A'/)
 720 FORMAT(//'   MAT',6X,'E',11X,'A',9X,'IZZ'/)
 730 FORMAT(//'   MAT',6X,'E',10X,'NU',10X,'BX',10X,'BY'/)
 740 FORMAT(//'   MAT',6X,'E',10X,'NU',10X,'BR'/)
 750 FORMAT(//'   MAT',6X,'G'/)
 760 FORMAT(I5,4E12.4)
 770 FORMAT(///' DISPLACEMENT BOUNDARY CONDITIONS'/)
 780 FORMAT(3('  NO NODE DOF   VALUE   ')/)
 790 FORMAT(3(1X,I4,I5,I4,E11.3,1X))
 800 FORMAT(///' FORCE BOUNDARY CONDITIONS'/)
 810 FORMAT(///' STRESS BOUNDARY CONDITIONS'/)
 820 FORMAT('  NO ELEM  M    N       SM          TM',
     *'        SN          TN'/)
 830 FORMAT(4I5,4E13.3)

     RETURN
     END



     SUBROUTINE ACTIVE
C ***** IDENTIFY ACTIVE EQUATIONS : SEE SECTION 7.4.1  *****
     COMMON/ELEM/IETYPE,NOE,IJ(1000,3),MAT(1000),NODN,NONE,NODE,NOST
     COMMON/NODE/NONS,XY(600,2),NOCO
     COMMON/EQNS/NODS,NOEQ,IBW,IDA(600,6),SS(1200,50),FS(1200)
     COMMON/DBCS/NOPD,IPD(50,2),PD(50)
     DO 10 NS=1,NONS
     DO 10 IDN=1,NODN
  10 IDA(NS,IDN)=1
     DO 20 IBC=1,NOPD
     IF(PD(IBC).NE.0.) GO TO 20
     NS=IPD(IBC,1)
     IDN=IPD(IBC,2)
     IDA(NS,IDN)=0
  20 CONTINUE
     NOEQ=0
     DO 30 NS=1,NONS
     DO 30 IDN=1,NODN
     IF(IDA(NS,IDN).EQ.0) GOTO 30
     NOEQ=NOEQ+1
     IDA(NS,IDN)=NOEQ
  30 CONTINUE
     RETURN
     END


     SUBROUTINE BWIDTH
C ***** CALCULATE SEMI-BANDWIDTH : SEE SECTION 7.4.2  *****
     COMMON/ELEM/IETYPE,NOE,IJ(1000,3),MAT(1000),NODN,NONE,NODE,NOST
     COMMON/EQNS/NODS,NOEQ,IBW,IDA(600,6),SS(1200,50),FS(1200)
```

```
      DIMENSION IRC(6)
      IBW=0
      DO 40 IE=1,NOE
      DO 10 NE=1,NONE
      NS=IJ(IE,NE)
      DO 10 IDN=1,NODN
      IDE=NODN*(NE-1)+IDN
   10 IRC(IDE)=IDA(NS,IDN)
      DO 30 IRE=1,NODE
      IRS=IRC(IRE)
      IF(IRS.EQ.0) GOTO 30
      DO 20 ICE=1,NODE
      ICS=IRC(ICE)-IRS+1
      IF(ICS.LT.1) GO TO 20
      IBW=MAX(IBW,ICS)
   20 CONTINUE
   30 CONTINUE
   40 CONTINUE
      RETURN
      END

      SUBROUTINE ASSEM
C ***** ASSEMBLE EQUILIBRIUM EQUATIONS : SEE SECTION 7.4.3 *****
      COMMON/ELEM/IETYPE,NOE,IJ(1000,3),MAT(1000),NODN,NONE,NODE,NOST
      COMMON/EQNS/NODS,NOEQ,IBW,IDA(600,6),SS(1200,50),FS(1200)
      COMMON/DBCS/NOPD,IPD(50,2),PD(50)
      COMMON/FBCS/NOPF,IPF(20,2),PF(20)
      DIMENSION IRC(6),SE(6,6),FE(6)
C ZERO ARRAYS
      DO 10 IRS=1,NOEQ
      FS(IRS)=0.
      DO 10 ICS=1,IBW
   10 SS(IRS,ICS)=0.
C SUM OVER ALL ELEMENTS IN SYSTEM
      DO 50 IE=1,NOE
      DO 15 IDE=1,NODE
   15 FE(IDE)=0.
      CALL ELEMNT(IE,SE,FE)
      DO 20 NE=1,NONE
      NS=IJ(IE,NE)
      DO 20 IDN=1,NODN
      IDE=NODN*(NE-1)+IDN
   20 IRC(IDE)=IDA(NS,IDN)
      DO 40 IRE=1,NODE
      IRS=IRC(IRE)
      IF(IRS.EQ.0) GOTO 40
      FS(IRS)=FS(IRS)+FE(IRE)
      DO 30 ICE=1,NODE
      ICS=IRC(ICE)-IRS+1
      IF(ICS.LT.1) GOTO 30
      SS(IRS,ICS)=SS(IRS,ICS)+SE(IRE,ICE)
   30 CONTINUE
   40 CONTINUE
   50 CONTINUE
C PRESCRIBED NODAL POINT FORCES
      DO 60 IBC=1,NOPF
      NS=IPF(IBC,1)
      IDN=IPF(IBC,2)
      IDS=IDA(NS,IDN)
   60 FS(IDS)=FS(IDS)+PF(IBC)
C DISPLACEMENT BOUNDARY CONDITIONS  : SEE SECTION 5.3.4
      IF(NOEQ.GT.(NODS-NOPD)) THEN
      DO 70 IBC=1,NOPD
      IF(PD(IBC).EQ.0.) GOTO 70
      NS=IPD(IBC,1)
```

```
          IDN=IPD(IBC,2)
          IDS=IDA(NS,IDN)
          SS(IDS,1)=SS(IDS,1)*1.E20
          FS(IDS)=PD(IBC)*SS(IDS,1)
       70 CONTINUE
          ENDIF
          RETURN
          END


          SUBROUTINE ELEMNT(IE,SE,FE)
C   *****  ACCESS ELEMENT LIBRARY FOR EQUILIBRIUM EQUATIONS  *****
          COMMON/ELEM/IETYPE,NOE,IJ(1000,3),MAT(1000),NODN,NONE,NODE,NOST
          DIMENSION SE(6,6),FE(6)
          GO TO (10,20,30,40,50,60),IETYPE
       10 CALL ELEM1(IE,SE,FE)
          GO TO 100
       20 CALL ELEM2(IE,SE,FE)
          GO TO 100
       30 CALL ELEM3(IE,SE,FE)
          GO TO 100
       40 CALL ELEM4(IE,SE,FE)
          GO TO 100
       50 CALL ELEM5(IE,SE,FE)
          GO TO 100
       60 CALL ELEM6(IE,SE,FE)
      100 RETURN
          END


          SUBROUTINE SOLVE
C   *****  SOLVE SET OF BANDED SYMMETRIC EQUILIBRIUM EQUATIONS   *****
C   *****        USING GAUSSIAN ELIMINATION : SEE SECTION 7.5.2      *****
          COMMON/EQNS/NODS,NOEQ,IBW,IDA(600,6),SS(1200,50),FS(1200)
          DIMENSION DS(1200)
C   STORE DIAGONAL COEFFICIENTS
          DO 10 IEQ=1,NOEQ
       10 DS(IEQ)=SS(IEQ,1)
C   FORWARD REDUCTION : SEE (7.7),(7.8)
          DO 50 I=1,NOEQ-1
          DO 40 IR=I+1,I+IBW-1
          IF(SS(I,IR-I+1).EQ.0.) GO TO 40
          R=SS(I,IR-I+1)/SS(I,1)
          DO 20 IC=IR,I+IBW-1
          IF(IC.GT.NOEQ) GO TO 30
          IF(SS(I,IC-I+1).EQ.0.) GO TO 20
          SS(IR,IC-IR+1)=SS(IR,IC-IR+1)-R*SS(I,IC-I+1)
       20 CONTINUE
       30 IF(FS(I).EQ.0.) GO TO 40
          FS(IR)=FS(IR)-R*FS(I)
       40 CONTINUE
          CN=SS(I+1,1)/DS(I+1)
          IF(CN.LT.1E-3) GOTO 80
       50 CONTINUE
C   BACK SUBSTITUTION : SEE (7.10),(7.11)
          FS(NOEQ)=FS(NOEQ)/SS(NOEQ,1)
          DO 70 IR=NOEQ-1,1,-1
          DO 60 IC=IR+1,IR+IBW-1
          IF(IC.GT.NOEQ) GO TO 70
          IF(SS(IR,IC-IR+1).EQ.0..OR.FS(IC).EQ.0.)GO TO 60
          FS(IR)=FS(IR)-SS(IR,IC-IR+1)*FS(IC)
       60 CONTINUE
       70 FS(IR)=FS(IR)/SS(IR,1)
          CALL RENUM
          RETURN
       80 WRITE(6,600)
      600 FORMAT(//'   ***** ILL-CONDITIONED EQUATIONS ***** ')
```

```
      STOP
      END

      SUBROUTINE RENUM
C ***** RENUMBER DISPLACEMENTS : SEE SECTION 7.4.1 *****
      COMMON/ELEM/IETYPE,NOE,IJ(1000,3),MAT(1000),NODN,NONE,NODE,NOST
      COMMON/EQNS/NODS,NOEQ,IBW,IDA(600,6),SS(1200,50),FS(1200)
      COMMON/NODE/NONS,XY(600,2),NOCO
      DO 10 NS=NONS,1,-1
      DO 10 IDN=NODN,1,-1
      IDS=NODN*(NS-1)+IDN
      IF(IDA(NS,IDN).EQ.0) THEN
      FS(IDS)=0.
      ELSE
      FS(IDS)=FS(IDA(NS,IDN))
      ENDIF
   10 CONTINUE
      RETURN
      END

      SUBROUTINE STRESS
C ***** ACCESS ELEMENT LIBRARY FOR STRESS CALCULATIONS *****
      COMMON/ELEM/IETYPE,NOE,IJ(1000,3),MAT(1000),NODN,NONE,NODE,NOST
      GO TO (10,20,30,40,50,60),IETYPE
   10 CALL STR1
      GO TO 100
   20 CALL STR2
      GO TO 100
   30 CALL STR3
      GO TO 100
   40 CALL STR4
      GO TO 100
   50 CALL STR5
      GO TO 100
   60 CALL STR6
  100 RETURN
      END


      SUBROUTINE OUTPUT
C ***** LIST RESULTS *****
      COMMON/ELEM/IETYPE,NOE,IJ(1000,3),MAT(1000),NODN,NONE,NODE,NOST
      COMMON/NODE/NONS,XY(600,2),NOCO
      COMMON/EQNS/NODS,NOEQ,IBW,IDA(600,6),SS(1200,50),FS(1200)
      WRITE(6,600)
C DISPLACEMENTS
      WRITE(6,610)
      GO TO (10,20,30,40,50,60) IETYPE
   10 WRITE(6,620)
      WRITE(6,630)(NS,(FS((NS-1)*NODN+IDN),IDN=1,NODN),NS=1,NONS)
      GO TO 70
   20 WRITE(6,640)
      WRITE(6,650)(NS,(FS((NS-1)*NODN+IDN),IDN=1,NODN),NS=1,NONS)
      GO TO 70
   30 WRITE(6,620)
      WRITE(6,630)(NS,(FS((NS-1)*NODN+IDN),IDN=1,NODN),NS=1,NONS)
      GO TO 70
   40 WRITE(6,660)
      WRITE(6,630)(NS,(FS((NS-1)*NODN+IDN),IDN=1,NODN),NS=1,NONS)
      GO TO 70
   50 WRITE(6,670)
      WRITE(6,680)(NS,(FS((NS-1)*NODN+IDN),IDN=1,NODN),NS=1,NONS)
      GO TO 70
   60 WRITE(6,690)
      WRITE(6,680)(NS,(FS((NS-1)*NODN+IDN),IDN=1,NODN),NS=1,NONS)
```

```
      70 CONTINUE
   C   STRESSES
         WRITE(6,700)
         GO TO (80,90,100,110,120,130) IETYPE
      80 WRITE(6,710)
         WRITE(6,720)(IE,(SS(IE,IST),IST=1,NOST),IE=1,NOE)
         GO TO 140
      90 WRITE(6,730)
         WRITE(6,740)(IE,(SS(IE,IST),IST=1,NOST),IE=1,NOE)
         GO TO 140
     100 WRITE(6,750)
         WRITE(6,760)(IE,(SS(IE,IST),IST=1,NOST),IE=1,NOE)
         GO TO 140
     110 WRITE(6,770)
         WRITE(6,780)(IE,(SS(IE,IST),IST=1,NOST),IE=1,NOE)
         GO TO 140
     120 WRITE(6,790)
         WRITE(6,800)(IE,(SS(IE,IST),IST=1,NOST),IE=1,NOE)
         WRITE(6,810)SS(1,3)
         GO TO 140
     130 WRITE(6,820)
         WRITE(6,800)(IE,(SS(IE,IST),IST=1,NOST),IE=1,NOE)
     140 CONTINUE
     600 FORMAT(///' PROBLEM SOLUTION'/
        *' ****************'//)
     610 FORMAT(' NODAL DISPLACEMENTS'//)
     620 FORMAT(3(' NODE',5X,'U',10X,'V',4X)/)
     630 FORMAT(3(1X,I3,2E11.3))
     640 FORMAT(2(' NODE',5X,'U',10X,'V',8X,'THETA',2X)/)
     650 FORMAT(2(1X,I3,3E11.3))
     660 FORMAT(3(' NODE',5X,'U',10X,'W',4X)/)
     670 FORMAT(4(' NODE',4X,'W/A',3X)/)
     680 FORMAT(4(1X,I3,E11.3))
     690 FORMAT(4(' NODE',4X,'V/R',3X)/)
     700 FORMAT(///' ELEMENT STRESSES'/)
     710 FORMAT(4(' ELEM',3X,'STRESS',2X)/)
     720 FORMAT(4(1X,I3,E11.3,1X))
     730 FORMAT(' ELEM',4X,'FI',9X,'VI',9X,'MI',9X,
        * 'FJ',9X,'VJ',9X,'MJ'/)
     740 FORMAT(I4,6E11.3)
     750 FORMAT(2(' ELEM',3X,'SIG-XX',5X,'SIG-YY',5X,'SIG-XY',2X)/)
     760 FORMAT(2(1X,I3,3E11.3,1X))
     770 FORMAT(' ELEM',3X,'SIG-RR',5X,'SIG-ZZ',5X,'SIG-RZ',5X,
        *'SIG-TT',2X)/)
     780 FORMAT(1X,I3,4E11.3)
     790 FORMAT(3(' ELEM',3X,'SIG-YZ',5X,'SIG-ZX',2X)/)
     800 FORMAT(3(1X,I3,2E11.3,1X))
     810 FORMAT(////' TORSIONAL STIFFNESS T/AG = ',E12.4)
     820 FORMAT(3(' ELEM',3X,'SIG-RT',5X,'SIG-TZ',2X)/)
         RETURN
         END

         SUBROUTINE ELEM1(IE,SE,FE)
   C ***** EQUILIBRIUM EQUATIONS FOR PIN-JOINTED BAR ELEMENT *****
   C *****                 SEE SECTION 8.2.1                  *****
         COMMON/ELEM/IETYPE,NOE,IJ(1000,3),MAT(1000),NODN,NONE,NODE,NOST
         COMMON/NODE/NONS,XY(600,2),NOCO
         COMMON/MATS/NOPROP,NOSET,EDAT(5,5)
         COMMON/SBCS/NOPS,IPS(50,3),PS(50,4),JPS(1000)
         DIMENSION SE(6,6),FE(6)
         I=IJ(IE,1)
         J=IJ(IE,2)
         ISET=MAT(IE)
         EM=EDAT(ISET,1)
         XSA=EDAT(ISET,2)
```

```
      FLEN=SQRT((XY(J,1)-XY(I,1))**2+(XY(J,2)-XY(I,2))**2)
      FACT=EM*XSA/FLEN
      C=(XY(J,1)-XY(I,1))/FLEN
      S=(XY(J,2)-XY(I,2))/FLEN
      SE(1,1)= C*C*FACT
      SE(1,2)= C*S*FACT
      SE(1,3)=-C*C*FACT
      SE(1,4)=-C*S*FACT
      SE(2,2)= S*S*FACT
      SE(2,3)=-C*S*FACT
      SE(2,4)=-S*S*FACT
      SE(3,3)= C*C*FACT
      SE(3,4)= C*S*FACT
      SE(4,4)= S*S*FACT
      DO 10 IR=2,4
      DO 10 IC=1,IR-1
   10 SE(IR,IC)=SE(IC,IR)
      RETURN
      END


      SUBROUTINE STR1
C  ***** AXIAL STRESS IN PLANE PIN-JOINTED BAR ELEMENT  *****
C  *****               SEE (8.5)                        *****
      COMMON/EQNS/NODS,NOEQ,IBW,IDA(600,6),SS(1200,50),FS(1200)
      COMMON/ELEM/IETYPE,NOE,IJ(1000,3),MAT(1000),NODN,NONE,NODE,NOST
      COMMON/NODE/NONS,XY(600,2),NOCO
      COMMON/MATS/NOPROP,NOSET,EDAT(5,5)
      DO 10 IE=1,NOE
      I=IJ(IE,1)
      J=IJ(IE,2)
      ISET=MAT(IE)
      EM=EDAT(ISET,1)
      XSA=EDAT(ISET,2)
      FLEN=SQRT((XY(J,1)-XY(I,1))**2+(XY(J,2)-XY(I,2))**2)
      FACT=EM*XSA/FLEN
      C=(XY(J,1)-XY(I,1))/FLEN
      S=(XY(J,2)-XY(I,2))/FLEN
      DL=(FS(2*J-1)*C+FS(2*J)*S)-(FS(2*I-1)*C+FS(2*I)*S)
   10 SS(IE,1)=EM*DL/FLEN
      RETURN
      END



      SUBROUTINE ELEM2(IE,SE,FE)
C  ***** EQUILIBRIUM EQUATIONS FOR RIGID-JOINTED BEAM ELEMENT  *****
C  *****                SEE SECTION 8.3.2                      *****
      COMMON/ELEM/IETYPE,NOE,IJ(1000,3),MAT(1000),NODN,NONE,NODE,NOST
      COMMON/NODE/NONS,XY(600,2),NOCO
      COMMON/MATS/NOPROP,NOSET,EDAT(5,5)
      COMMON/SBCS/NOPS,IPS(50,3),PS(50,4),JPS(1000)
      DIMENSION SE(6,6),FE(6)
C ELEMENT STIFFNESS MATRIX : SEE (8.16)
      I=IJ(IE,1)
      J=IJ(IE,2)
      ISET=MAT(IE)
      EM=EDAT(ISET,1)
      XSA=EDAT(ISET,2)
      FIZZ=EDAT(ISET,3)
      FLEN=SQRT((XY(J,1)-XY(I,1))**2+(XY(J,2)-XY(I,2))**2)
      E=EM*XSA/FLEN
      F=12.*EM*FIZZ/FLEN**3
      G=6.*EM*FIZZ/FLEN**2
      H=4.*EM*FIZZ/FLEN
      C=(XY(J,1)-XY(I,1))/FLEN
      S=(XY(J,2)-XY(I,2))/FLEN
```

```
              SE(1,1)= C*C*E+S*S*F
              SE(1,2)= C*S*(E-F)
              SE(1,3)= S*G
              SE(1,4)=-C*C*E-S*S*F
              SE(1,5)= C*S*(-E+F)
              SE(1,6)= S*G
              SE(2,2)= S*S*E+C*C*F
              SE(2,3)=-C*G
              SE(2,4)= C*S*(-E+F)
              SE(2,5)=-S*S*E-C*C*F
              SE(2,6)=-C*G
              SE(3,3)= H
              SE(3,4)=-S*G
              SE(3,5)= C*G
              SE(3,6)= .5*H
              SE(4,4)= C*C*E+S*S*F
              SE(4,5)= C*S*(E-F)
              SE(4,6)=-S*G
              SE(5,5)= S*S*E+C*C*F
              SE(5,6)= C*G
              SE(6,6)= H
              DO 10 IR=2,6
              DO 10 IC=1,IR-1
           10 SE(IR,IC)=SE(IC,IR)
C     LINEARLY VARYING DISTRIBUTED LOAD : SEE (8.17)
              IBC=JPS(IE)
              IF(IBC.EQ.0) RETURN
              M=IPS(IBC,2)
              N=IPS(IBC,3)
              IF(I.EQ.M) THEN
              MM=1
              NN=2
              ELSE
              MM=2
              NN=1
              ENDIF
              FE(3*MM-2)=  S*(7.*FLEN/20.*PS(IBC,1)+  3.*FLEN/20.*PS(IBC,3))
              FE(3*MM-1)= -C*(7.*FLEN/20.*PS(IBC,1)+  3.*FLEN/20.*PS(IBC,3))
              FE(3*MM)=      FLEN*FLEN/20.*PS(IBC,1)+FLEN*FLEN/30.*PS(IBC,3)
              FE(3*NN-2)=  S*(3.*FLEN/20.*PS(IBC,1)+  7.*FLEN/20.*PS(IBC,3))
              FE(3*NN-1)= -C*(3.*FLEN/20.*PS(IBC,1)+  7.*FLEN/20.*PS(IBC,3))
              FE(3*NN)=     -FLEN*FLEN/30.*PS(IBC,1)-FLEN*FLEN/20.*PS(IBC,3)
              RETURN
              END


              SUBROUTINE STR2
C     *****   STRESS RESULTANTS IN RIGID-JOINTED BEAM ELEMENT   *****
C     *****                    SEE (8.18)                       *****
              COMMON/EQNS/NODS,NOEQ,IBW,IDA(600,6),SS(1200,50),FS(1200)
              COMMON/ELEM/IETYPE,NOE,IJ(1000,3),MAT(1000),NODN,NONE,NODE,NOST
              COMMON/NODE/NONS,XY(600,2),NOCO
              COMMON/MATS/NOPROP,NOSET,EDAT(5,5)
              DIMENSION SE(6,6),FE(6)
              DO 20 IE=1,NOE
              I=IJ(IE,1)
              J=IJ(IE,2)
              ISET=MAT(IE)
              EM=EDAT(ISET,1)
              XSA=EDAT(ISET,2)
              FIZZ=EDAT(ISET,3)
              FLEN=SQRT((XY(J,1)-XY(I,1))**2+(XY(J,2)-XY(I,2))**2)
              E=EM*XSA/FLEN
              F=12.*EM*FIZZ/FLEN**3
              G=6.*EM*FIZZ/FLEN**2
              H=4.*EM*FIZZ/FLEN
```

```
      C=(XY(J,1)-XY(I,1))/FLEN
      S=(XY(J,2)-XY(I,2))/FLEN
      SE(1,1)= C*E
      SE(1,2)= S*E
      SE(1,3)= 0.
      SE(1,4)=-C*E
      SE(1,5)=-S*E
      SE(1,6)= 0.
      SE(2,1)= S*F
      SE(2,2)=-C*F
      SE(2,3)= G
      SE(2,4)=-S*F
      SE(2,5)= C*F
      SE(2,6)= G
      SE(3,1)= S*G
      SE(3,2)=-C*G
      SE(3,3)= H
      SE(3,4)=-S*G
      SE(3,5)= C*G
      SE(3,6)= .5*H
      SE(4,1)=-C*E
      SE(4,2)=-S*E
      SE(4,3)= 0.
      SE(4,4)= C*E
      SE(4,5)= S*E
      SE(4,6)= 0.
      SE(5,1)=-S*F
      SE(5,2)= C*F
      SE(5,3)=-G
      SE(5,4)= S*F
      SE(5,5)=-C*F
      SE(5,6)=-G
      SE(6,1)= S*G
      SE(6,2)=-C*G
      SE(6,3)= .5*H
      SE(6,4)=-S*G
      SE(6,5)= C*G
      SE(6,6)= H
      FE(1)=FS(3*I-2)
      FE(2)=FS(3*I-1)
      FE(3)=FS(3*I)
      FE(4)=FS(3*J-2)
      FE(5)=FS(3*J-1)
      FE(6)=FS(3*J)
      DO 10 IR=1,6
      SS(IE,IR)=0.
      DO 10 IC=1,6
   10 SS(IE,IR)=SS(IE,IR)+SE(IR,IC)*FE(IC)
   20 CONTINUE
      RETURN
      END


      SUBROUTINE ELEM3(IE,SE,FE)
C     *****  EQUILIBRIUM EQUATIONS FOR PLANE CST ELEMENT  *****
C     *****                 SEE SECTION 9.4               *****
      COMMON/ELEM/IETYPE,NOE,IJ(1000,3),MAT(1000),NODN,NONE,NODE,NOST
      COMMON/NODE/NONS,XY(600,2),NOCO
      COMMON/MATS/NOPROP,NOSET,EDAT(5,5)
      COMMON/SBCS/NOPS,IPS(50,3),PS(50,4),JPS(1000)
      DIMENSION B(4,6),D(4,4),SE(6,6),FE(6),S(4,6)
C ELEMENT STIFFNESS MATRIX : SEE SECTION 9.4.2
      I=IJ(IE,1)
      J=IJ(IE,2)
      K=IJ(IE,3)
```

```
                ISET=MAT(IE)
                EM=EDAT(ISET,1)
                PR=EDAT(ISET,2)
                AI=XY(J,1)*XY(K,2)-XY(K,1)*XY(J,2)
                AJ=XY(K,1)*XY(I,2)-XY(I,1)*XY(K,2)
                AK=XY(I,1)*XY(J,2)-XY(J,1)*XY(I,2)
                AREA=.5*(AI+AJ+AK)
                BI=XY(J,2)-XY(K,2)
                BJ=XY(K,2)-XY(I,2)
                BK=XY(I,2)-XY(J,2)
                CI=XY(K,1)-XY(J,1)
                CJ=XY(I,1)-XY(K,1)
                CK=XY(J,1)-XY(I,1)
C    MATRIX [B] : SEE (9.21)
                DO 10 IR=1,NOST
                DO 10 IC=1,NODE
           10   B(IR,IC)=0.
                B(1,1)=BI
                B(1,3)=BJ
                B(1,5)=BK
                B(2,2)=CI
                B(2,4)=CJ
                B(2,6)=CK
                B(3,1)=CI
                B(3,2)=BI
                B(3,3)=CJ
                B(3,4)=BJ
                B(3,5)=CK
                B(3,6)=BK
C    MATRIX [D] : SEE (9.22)
                C=.25*EM/((1.-PR*PR)*AREA)
                DO 20 IR=1,NOST
                DO 20 IC=1,NOST
           20   D(IR,IC)=0.
                D(1,1)=C
                D(1,2)=C*PR
                D(2,1)=C*PR
                D(2,2)=C
                D(3,3)=C*.5*(1.-PR)
C    MATRIX PRODUCT [S]=[D][B]
                DO 30 IR=1,NOST
                DO 30 IC=1,NODE
                S(IR,IC)=0.
                DO 30 KK=1,NOST
           30   S(IR,IC)=S(IR,IC)+D(IR,KK)*B(KK,IC)
C    ELEMENT STIFFNESS MATRIX [K]=[B]T[S]xVOL : SEE (9.23)
                DO 40 IR=1,NODE
                DO 40 IC=IR,NODE
                SE(IR,IC)=0.
                DO 40 KK=1,NOST
           40   SE(IR,IC)=SE(IR,IC)+B(KK,IR)*S(KK,IC)
                DO 50 IR=2,NODE
                DO 50 IC=1,IR-1
           50   SE(IR,IC)=SE(IC,IR)
C    LINEARLY VARYING BOUNDARY TRACTIONS : SEE SECTION 9.4.3
                IF(JPS(IE).NE.0) THEN
                IBC=JPS(IE)
                M=IPS(IBC,2)
                N=IPS(IBC,3)
                DO 60 NE=1,NONE
                IF(IJ(IE,NE).EQ.M) MM=NE
                IF(IJ(IE,NE).EQ.N) NN=NE
           60   CONTINUE
                DX=XY(M,1)-XY(N,1)
                DY=XY(N,2)-XY(M,2)
```

```
            DL=SQRT(DX*DX+DY*DY)
            SINE=DX/DL
            COSINE=DY/DL
C     TRACTION COMPONENTS : SEE (9.26)
            TXM=PS(IBC,1)*COSINE-PS(IBC,2)*SINE
            TXN=PS(IBC,3)*COSINE-PS(IBC,4)*SINE
            TYM=PS(IBC,1)*SINE+PS(IBC,2)*COSINE
            TYN=PS(IBC,3)*SINE+PS(IBC,4)*COSINE
C     EQUIVALENT NODAL FORCES : SEE (9.29)
            FE(2*MM-1)=FE(2*MM-1)+DL*TXM/3.+DL*TXN/6.
            FE(2*MM)  =FE(2*MM)  +DL*TYM/3.+DL*TYN/6.
            FE(2*NN-1)=FE(2*NN-1)+DL*TXM/6.+DL*TXN/3.
            FE(2*NN)  =FE(2*NN)  +DL*TYM/6.+DL*TYN/3.
            ENDIF
C     UNIFORM BODY FORCES : SEE (9.32)
            BX=EDAT(ISET,3)
            BY=EDAT(ISET,4)
            IF(BX.NE.0.) THEN
            DO 70 II=1,3
   70    FE(2*II-1)=FE(2*II-1)+AREA*BX/3.
            ENDIF
            IF(BY.NE.0.) THEN
            DO 80 II=1,3
   80    FE(2*II)=FE(2*II)+AREA*BY/3.
            ENDIF
            RETURN
            END

            SUBROUTINE STR3
C     *****  STRESSES IN PLANE CST ELEMENT : SEE SECTION 9.4.5  *****
            COMMON/EQNS/NODS,NOEQ,IBW,IDA(600,6),SS(1200,50),FS(1200)
            COMMON/ELEM/IETYPE,NOE,IJ(1000,3),MAT(1000),NODN,NONE,NODE,NOST
            COMMON/NODE/NONS,XY(600,2),NOCO
            COMMON/MATS/NOPROP,NOSET,EDAT(5,5)
            DIMENSION STR(6)
            DO 10 IE=1,NOE
            I=IJ(IE,1)
            J=IJ(IE,2)
            K=IJ(IE,3)
            ISET=MAT(IE)
            EM=EDAT(ISET,1)
            PR=EDAT(ISET,2)
            AI=XY(J,1)*XY(K,2)-XY(K,1)*XY(J,2)
            AJ=XY(K,1)*XY(I,2)-XY(I,1)*XY(K,2)
            AK=XY(I,1)*XY(J,2)-XY(J,1)*XY(I,2)
            AREA=.5*(AI+AJ+AK)
            BI=XY(J,2)-XY(K,2)
            BJ=XY(K,2)-XY(I,2)
            BK=XY(I,2)-XY(J,2)
            CI=XY(K,1)-XY(J,1)
            CJ=XY(I,1)-XY(K,1)
            CK=XY(J,1)-XY(I,1)
C     ELEMENT STRAINS : SEE (9.21)
            EXX=.5*(BI*FS(2*I-1)+BJ*FS(2*J-1)+BK*FS(2*K-1))/AREA
            EYY=.5*(CI*FS(2*I)   +CJ*FS(2*J)   +CK*FS(2*K))/AREA
            GXY=.5*(CI*FS(2*I-1)+BI*FS(2*I)+CJ*FS(2*J-1)+BJ*FS(2*J)+
          *  CK*FS(2*K-1)+BK*FS(2*K))/AREA
C     ELEMENT STRESSES : SEE (9.22)
            C=EM/(1.-PR*PR)
            SS(IE,1)=C*(EXX+PR*EYY)
            SS(IE,2)=C*(EYY+PR*EXX)
   10    SS(IE,3)=.5*(1.-PR)*C*GXY
            RETURN
            END
```

```
         SUBROUTINE ELEM4(IE,SE,FE)
C    *****   EQUILIBRIUM EQUATIONS FOR AXISYMMETRIC SOLID ELEMENT   *****
C    *****                    SEE SECTION 10.3                      *****
         COMMON/ELEM/IETYPE,NOE,IJ(1000,3),MAT(1000),NODN,NONE,NODE,NOST
         COMMON/NODE/NONS,XY(600,2),NOCO
         COMMON/MATS/NOPROP,NOSET,EDAT(5,5)
         COMMON/SBCS/NOPS,IPS(50,3),PS(50,4),JPS(1000)
         DIMENSION B(4,6),D(4,4),SE(6,6),FE(6),S(4,6)
         PI=4.*ATAN(1.)
C ELEMENT STIFFNESS MATRIX : SEE SECTION 10.3.2
         I=IJ(IE,1)
         J=IJ(IE,2)
         K=IJ(IE,3)
         ISET=MAT(IE)
         EM=EDAT(ISET,1)
         PR=EDAT(ISET,2)
         RC=(XY(I,1)+XY(J,1)+XY(K,1))/3.
         ZC=(XY(I,2)+XY(J,2)+XY(K,2))/3.
         AI=XY(J,1)*XY(K,2)-XY(K,1)*XY(J,2)
         AJ=XY(K,1)*XY(I,2)-XY(I,1)*XY(K,2)
         AK=XY(I,1)*XY(J,2)-XY(J,1)*XY(I,2)
         AREA=.5*(AI+AJ+AK)
         BI=XY(J,2)-XY(K,2)
         BJ=XY(K,2)-XY(I,2)
         BK=XY(I,2)-XY(J,2)
         CI=XY(K,1)-XY(J,1)
         CJ=XY(I,1)-XY(K,1)
         CK=XY(J,1)-XY(I,1)
C MATRIX [B] : SEE (10.14)
         DO 10 IR=1,NOST
         DO 10 IC= 1,NODE
      10 B(IR,IC)=0.
         B(1,1)=BI
         B(1,3)=BJ
         B(1,5)=BK
         B(2,2)=CI
         B(2,4)=CJ
         B(2,6)=CK
         B(3,1)=CI
         B(3,2)=BI
         B(3,3)=CJ
         B(3,4)=BJ
         B(3,5)=CK
         B(3,6)=BK
         B(4,1)=AI/RC+BI+CI*ZC/RC
         B(4,3)=AJ/RC+BJ+CJ*ZC/RC
         B(4,5)=AK/RC+BK+CK*ZC/RC
C MATRIX [D] : SEE (10.15)
         C=EM/((1.+PR)*(1.-2.*PR))*PI*RC/(2.*AREA)
         DO 20 IR=1,NOST
         DO 20 IC=1,NOST
      20 D(IR,IC)=0.
         D(1,1)=C*(1.-PR)
         D(1,2)=C*PR
         D(1,4)=C*PR
         D(2,1)=C*PR
         D(2,2)=C*(1.-PR)
         D(2,4)=C*PR
         D(3,3)=C*.5*(1.-2.*PR)
         D(4,1)=C*PR
         D(4,2)=C*PR
         D(4,4)=C*(1.-PR)
C MATRIX PRODUCT [S]=[D][B]
         DO 30 IR=1,NOST
         DO 30 IC=1,NODE
```

```
       S(IR,IC)=0.
       DO 30 KK=1,NOST
    30 S(IR,IC)=S(IR,IC)+D(IR,KK)*B(KK,IC)
C  ELEMENT STIFFNESS MATRIX [K]=[B]T[S] VOLUME : SEE (10.18)
       DO 40 IR=1,NODE
       DO 40 IC=IR,NODE
       SE(IR,IC)=0.
       DO 40 KK=1,NOST
    40 SE(IR,IC)=SE(IR,IC)+B(KK,IR)*S(KK,IC)
       DO 50 IR=2,NODE
       DO 50 IC=1,IR-1
    50 SE(IR,IC)=SE(IC,IR)
C  LINEARLY VARYING BOUNDARY TRACTIONS : SEE SECTION 10.3.3
       IF(JPS(IE).NE.0) THEN
       IBC=JPS(IE)
       M=IPS(IBC,2)
       N=IPS(IBC,3)
       DO 60 NE=1,NONE
       IF(IJ(IE,NE).EQ.M) MM=NE
       IF(IJ(IE,NE).EQ.N) NN=NE
    60 CONTINUE
       DR=XY(M,1)-XY(N,1)
       DZ=XY(N,2)-XY(M,2)
       DL=SQRT(DR*DR+DZ*DZ)
       SINE=DR/DL
       COSINE=DZ/DL
       RM=XY(M,1)
       RN=XY(N,1)
C  TRACTION COMPONENTS : SEE (10.19)
       TRM=PS(IBC,1)*COSINE-PS(IBC,2)*SINE
       TRN=PS(IBC,3)*COSINE-PS(IBC,4)*SINE
       TZM=PS(IBC,1)*SINE+PS(IBC,2)*COSINE
       TZN=PS(IBC,3)*SINE+PS(IBC,4)*COSINE
C  EQUIVALENT NODAL FORCES : SEE (10.21)
       FE(2*MM-1)=FE(2*MM-1)+2.*PI*DL*((RM/4.+ RN/12.)*TRM+
      *(RM/12.+RN/12.)*TRN)
       FE(2*MM)  =FE(2*MM)   +2.*PI*DL*((RM/4.+ RN/12.)*TZM+
      *(RM/12.+RN/12.)*TZN)
       FE(2*NN-1)=FE(2*NN-1)+2.*PI*DL*((RM/12.+RN/12.)*TRM+
      *(RM/12.+ RN/4.)*TRN)
       FE(2*NN)  =FE(2*NN)   +2.*PI*DL*((RM/12.+RN/12.)*TZM+
      *(RM/12.+ RN/4.)*TZN)
       ENDIF
C  ROTATIONAL BODY FORCES : SEE SECTION 10.3.4
       BR=EDAT(ISET,3)
       IF(BR.NE.0.) THEN
       RI=XY(I,1)-RC
       RJ=XY(J,1)-RC
       RK=XY(K,1)-RC
       SUMSQ=RI*RI+RJ*RJ+RK*RK
       FE(1)=FE(1)+2.*PI*BR*AREA*RC*RC/3.
       FE(3)=FE(3)+2.*PI*BR*AREA*RC*RC/3.
       FE(5)=FE(5)+2.*PI*BR*AREA*RC*RC/3.
       ENDIF
       RETURN
       END

       SUBROUTINE STR4
C  *****   STRESSES IN AXISYMMETRIC SOLID ELEMENT   *****
C  *****            SEE SECTION 10.3.5              *****
       COMMON/EQNS/NODS,NOEQ,IBW,IDA(600,6),SS(1200,50),FS(1200)
       COMMON/ELEM/IETYPE,NOE,IJ(1000,3),MAT(1000),NODN,NONE,NODE,NOST
       COMMON/MATS/NOPROP,NOSET,EDAT(5,5)
       COMMON/NODE/NONS,XY(600,2),NOCO
       DO 10 IE=1,NOE
```

```
      I=IJ(IE,1)
      J=IJ(IE,2)
      K=IJ(IE,3)
      ISET=MAT(IE)
      EM=EDAT(ISET,1)
      PR=EDAT(ISET,2)
      RC=(XY(I,1)+XY(J,1)+XY(K,1))/3.
      ZC=(XY(I,2)+XY(J,2)+XY(K,2))/3.
      AI=XY(J,1)*XY(K,2)-XY(K,1)*XY(J,2)
      AJ=XY(K,1)*XY(I,2)-XY(I,1)*XY(K,2)
      AK=XY(I,1)*XY(J,2)-XY(J,1)*XY(I,2)
      AREA=.5*(AI+AJ+AK)
      BI=XY(J,2)-XY(K,2)
      BJ=XY(K,2)-XY(I,2)
      BK=XY(I,2)-XY(J,2)
      CI=XY(K,1)-XY(J,1)
      CJ=XY(I,1)-XY(K,1)
      CK=XY(J,1)-XY(I,1)
C  ELEMENT STRAINS : SEE (10.14)
      ERR=.5*(BI*FS(2*I-1)+BJ*FS(2*J-1)+BK*FS(2*K-1))/AREA
      EZZ=.5*(CI*FS(2*I)+CJ*FS(2*J)+CK*FS(2*K))/AREA
      GRZ=.5*(CI*FS(2*I-1)+BI*FS(2*I)+CJ*FS(2*J-1)+BJ*FS(2*J)+
     * CK*FS(2*K-1)+BK*FS(2*K))/AREA
      ETT=.5*((AI/RC+BI+CI*ZC/RC)*FS(2*I-1)+(AJ/RC+BJ+CJ*ZC/RC)*
     * FS(2*J-1)+(AK/RC+BK+CK*ZC/RC)*FS(2*K-1))/AREA
C  ELEMENT STRESSES : SEE (10.15)
      C=EM/((1.+PR)*(1.-2.*PR))
      SS(IE,1)=C*((1.-PR)*ERR+PR*EZZ+PR*ETT)
      SS(IE,2)=C*(PR*ERR+(1.-PR)*EZZ+PR*ETT)
      SS(IE,3)=C*.5*(1.-2.*PR)*GRZ
   10 SS(IE,4)=C*(PR*ERR+PR*EZZ+(1.-PR)*ETT)
      RETURN
      END



      SUBROUTINE ELEM5(IE,SE,FE)
C  *****   EQUILIBRIUM EQUATIONS FOR NONCIRCULAR SHAFT ELEMENT   *****
C  *****            SEE SECTION 11.2.2                           *****
      COMMON/ELEM/IETYPE,NOE,IJ(1000,3),MAT(1000),NODN,NONE,NODE,NOST
      COMMON/NODE/NONS,XY(600,2),NOCO
      COMMON/MATS/NOPROP,NOSET,EDAT(5,5)
      COMMON/SBCS/NOPS,IPS(50,3),PS(50,4),JPS(1000)
      DIMENSION SE(6,6),FE(6)
C  ELEMENT STIFFNESS MATRIX : SEE (11.25)
      I=IJ(IE,1)
      J=IJ(IE,2)
      K=IJ(IE,3)
      AI=XY(J,1)*XY(K,2)-XY(K,1)*XY(J,2)
      AJ=XY(K,1)*XY(I,2)-XY(I,1)*XY(K,2)
      AK=XY(I,1)*XY(J,2)-XY(J,1)*XY(I,2)
      AREA=.5*(AI+AJ+AK)
      BI=XY(J,2)-XY(K,2)
      BJ=XY(K,2)-XY(I,2)
      BK=XY(I,2)-XY(J,2)
      CI=XY(K,1)-XY(J,1)
      CJ=XY(I,1)-XY(K,1)
      CK=XY(J,1)-XY(I,1)
      SE(1,1)=(BI*BI+CI*CI)/(4.*AREA)
      SE(1,2)=(BI*BJ+CI*CJ)/(4.*AREA)
      SE(1,3)=(BI*BK+CI*CK)/(4.*AREA)
      SE(2,2)=(BJ*BJ+CJ*CJ)/(4.*AREA)
      SE(2,3)=(BJ*BK+CJ*CK)/(4.*AREA)
      SE(3,3)=(BK*BK+CK*CK)/(4.*AREA)
      SE(2,1)=SE(1,2)
      SE(3,1)=SE(1,3)
```

```
        SE(3,2)=SE(2,3)
C   MATRIX [FE] : SEE (11.26)
        XC=(XY(I,1)+XY(J,1)+XY(K,1))/3.
        YC=(XY(I,2)+XY(J,2)+XY(K,2))/3.
        FE(1)=(YC*BI-XC*CI)/2.
        FE(2)=(YC*BJ-XC*CJ)/2.
        FE(3)=(YC*BK-XC*CK)/2.
        RETURN
        END


        SUBROUTINE STR5
C   *****   STRESSES AND TORSIONAL CONSTANT FOR NONCIRCULAR  *****
C   *****         SHAFT ELEMENT : SEE SECTION 11.2.2         *****
        COMMON/EQNS/NODS,NOEQ,IBW,IDA(600,6),SS(1200,50),FS(1200)
        COMMON/ELEM/IETYPE,NOE,IJ(1000,3),MAT(1000),NODN,NONE,NODE,NOST
        COMMON/NODE/NONS,XY(600,2),NOCO
        COMMON/MATS/NOPROP,NOSET,EDAT(5,5)
        SS(1,3)=0.
        DO 10 IE=1,NOE
        I=IJ(IE,1)
        J=IJ(IE,2)
        K=IJ(IE,3)
        XI=XY(I,1)
        XJ=XY(J,1)
        XK=XY(K,1)
        YI=XY(I,2)
        YJ=XY(J,2)
        YK=XY(K,2)
        XC=(XI+XJ+XK)/3.
        YC=(YI+YJ+YK)/3.
        AI=XJ*YK-XK*YJ
        AJ=XK*YI-XI*YK
        AK=XI*YJ-XJ*YI
        AREA=.5*(AI+AJ+AK)
        BI=YJ-YK
        BJ=YK-YI
        BK=YI-YJ
        CI=XK-XJ
        CJ=XI-XK
        CK=XJ-XI
C   TORSIONAL CONSTANT : SEE (11.28)
        F1=((XC*CI-YC*BI)*FS(I)+(XC*CJ-YC*BJ)*FS(J)+
     *   (XC*CK-YC*BK)*FS(K))/2.
        F2=AREA/6.*(XI*XI+XJ*XJ+XK*XK+YI*YI+YJ*YJ+YK*YK+XI*XJ+XJ*XK+
     *   XK*XI+YI*YJ+YJ*YK+YK*YI)
        SS(1,3)=SS(1,3)+F1+F2
C   ELEMENT STRESSES : SEE (11.27)
        SS(IE,1)=1./(2.*AREA)*(CI*FS(I)+CJ*FS(J)+CK*FS(K))+XC
 10     SS(IE,2)=1./(2.*AREA)*(BI*FS(I)+BJ*FS(J)+BK*FS(K))-YC
        RETURN
        END


        SUBROUTINE ELEM6(IE,SE,FE)
C   *****   EQUILIBRIUM EQUATIONS FOR AXISYMMETRIC SHAFT ELEMENT   *****
C   *****                  SEE SECTION 11.3.2                      *****
        COMMON/ELEM/IETYPE,NOE,IJ(1000,3),MAT(1000),NODN,NONE,NODE,NOST
        COMMON/NODE/NONS,XY(600,2),NOCO
        COMMON/MATS/NOPROP,NOSET,EDAT(5,5)
        COMMON/SBCS/NOPS,IPS(50,3),PS(50,4),JPS(1000)
        DIMENSION SE(6,6),FE(6)
C   ELEMENT STIFFNESS MATRIX : SEE (11.57)
        I=IJ(IE,1)
        J=IJ(IE,2)
        K=IJ(IE,3)
```

```
      RI=XY(I,1)
      RJ=XY(J,1)
      RK=XY(K,1)
      ZI=XY(I,2)
      ZJ=XY(J,2)
      ZK=XY(K,2)
      ISET=MAT(IE)
      SM=EDAT(ISET,1)
      AI=RJ*ZK-RK*ZJ
      AJ=RK*ZI-RI*ZK
      AK=RI*ZJ-RJ*ZI
      AREA=.5*(AI+AJ+AK)
      BI=ZJ-ZK
      BJ=ZK-ZI
      BK=ZI-ZJ
      CI=RK-RJ
      CJ=RI-RK
      CK=RJ-RI
      FI=.1*(RI*RI*RI+RJ*RJ*RJ+RK*RK*RK+RI*RJ*RK+RI*RI*RJ+
     * RJ*RJ*RK+RK*RK*RI+RI*RI*RK+RJ*RJ*RI+RK*RK*RJ)
      SE(1,1)=FI*SM/(4.*AREA)*(BI*BI+CI*CI)
      SE(1,2)=FI*SM/(4.*AREA)*(BI*BJ+CI*CJ)
      SE(1,3)=FI*SM/(4.*AREA)*(BI*BK+CI*CK)
      SE(2,2)=FI*SM/(4.*AREA)*(BJ*BJ+CJ*CJ)
      SE(2,3)=FI*SM/(4.*AREA)*(BJ*BK+CJ*CK)
      SE(3,3)=FI*SM/(4.*AREA)*(BK*BK+CK*CK)
      SE(2,1)=SE(1,2)
      SE(3,1)=SE(1,3)
      SE(3,2)=SE(2,3)
C  LINEARLY VARYING TRACTIONS : SEE (11.60)
      IBC=JPS(IE)
      IF(IBC.NE.0) THEN
      M=IPS(IBC,2)
      N=IPS(IBC,3)
      DO 10 NE=1,NONE
      IF(IJ(IE,NE).EQ.M) MM=NE
      IF(IJ(IE,NE).EQ.N) NN=NE
   10 CONTINUE
      DR=XY(M,1)-XY(N,1)
      DZ=XY(M,2)-XY(N,2)
      DL=SQRT(DR*DR+DZ*DZ)
      RM=XY(M,1)
      RN=XY(N,1)
      TM=PS(IBC,1)
      TN=PS(IBC,3)
      FE(MM)=FE(MM)-DL/60.*((12.*RM*RM+6.*RM*RN+2.*RN*RN)*TM+
     * (3.*RM*RM+4.*RM*RN+3.*RN*RN)*TN)
      FE(NN)=FE(NN)-DL/60.*((3.*RM*RM+4.*RM*RN+3.*RN*RN)*TM+
     * (2.*RM*RM+6.*RM*RN+12.*RN*RN)*TN)
      ENDIF
      RETURN
      END

      SUBROUTINE STR6
C  *****   STRESSES IN AXISYMMETRIC SHAFT ELEMENT : SEE (11.61)   *****
      COMMON/EQNS/NODS,NOEQ,IBW,IDA(600,6),SS(1200,50),FS(1200)
      COMMON/ELEM/IETYPE,NOE,IJ(1000,3),MAT(1000),NODN,NONE,NODE,NOST
      COMMON/NODE/NONS,XY(600,2),NOCO
      COMMON/MATS/NOPROP,NOSET,EDAT(5,5)
      DO 10 IE=1,NOE
      I=IJ(IE,1)
      J=IJ(IE,2)
      K=IJ(IE,3)
      ISET=MAT(IE)
      SM=EDAT(ISET,1)
```

```
      AI=XY(J,1)*XY(K,2)-XY(K,1)*XY(J,2)
      AJ=XY(K,1)*XY(I,2)-XY(I,1)*XY(K,2)
      AK=XY(I,1)*XY(J,2)-XY(J,1)*XY(I,2)
      AREA=.5*(AI+AJ+AK)
      BI=XY(J,2)-XY(K,2)
      BJ=XY(K,2)-XY(I,2)
      BK=XY(I,2)-XY(J,2)
      CI=XY(K,1)-XY(J,1)
      CJ=XY(I,1)-XY(K,1)
      CK=XY(J,1)-XY(I,1)
      RC=(XY(I,1)+XY(J,1)+XY(K,1))/3.
      ZC=(XY(I,2)+XY(J,2)+XY(K,2))/3.
      SS(IE,1)=SM*RC*(BI*FS(I)+BJ*FS(J)+BK*FS(K))/(2.*AREA)
   10 SS(IE,2)=SM*RC*(CI*FS(I)+CJ*FS(J)+CK*FS(K))/(2.*AREA)
      RETURN
      END
```

Further Reading

Alexander, J. M., *Strength of Materials*, Vol. I, Ellis Horwood (1981).

Bathe, K.-J., and Wilson, E. L., *Numerical Methods in Finite Element Analysis*, Prentice-Hall (1976).

Boresi, A. P., and Sidebottom, O. M., *Advanced Mechanics of Materials*, 4th edn, John Wiley (1985).

Broek, D., *Elementary Engineering Fracture Mechanics*, Noordhoff (1974).

Cheung, Y. K., and Yeo, M. F., *A Practical Introduction to Finite Element Analysis*, Pitman (1979).

Coker, E. G., and Filon, L. N. G., *A Treatise on Photoelasticity*, Cambridge University Press (1957).

Cook, R. D., *Concepts and Applications of Finite Element Analysis*, John Wiley (1974).

Dawe, D. J., *Matrix and Finite Element Displacement Analysis of Structures*, Clarendon (1984).

Desai, C. S. and Abel, J. F., *Introduction to the Finite Element Method*, Van Nostrand Reinhold (1972).

Dugdale, D. S., and Ruiz, C., *Elasticity for Engineers*, McGraw-Hill (1971).

Fenner, R. T., *Engineering Theory of Elasticity*, Ellis Horwood (1986).

Gallagher, R. H., *Finite Element Analysis Fundamentals*, Prentice-Hall (1975).

Guerney, R., Photoelastic study of centrifugal stresses in a single wheel and hub, *Proc. Soc. Exp. Stress Anal.*, **18**, 1 (1961).

Herrmann, L. R., Elastic torsional analysis of irregular shapes, *J. Eng. Mech. Div.*, *Am. Soc. Civ. Eng.*, **91**, EM6, 11 (1965).

Hetenyi, M., *Handbook of Experimental Stress Analysis*, Chapman and Hall (1950).

Hinton, E., and Owen, D. R. J., *Finite Element Programming*, Academic Press (1977).

Huebner, K. H., *The Finite Element Method for Engineers*, John Wiley (1975).

Irons, B., and Ahmad, S., *Techniques of Finite Element Analysis*, Ellis Horwood (1980).

Irons, B., and Shrive, N., *Finite Element Primer*, Ellis Horwood (1983).

Knott, J. F., *Fundamentals of Engineering Fracture Mechanics*, Butterworths (1973).

Livesley, R. K., *Matrix Methods of Structural Analysis*, Pergamon (1975).

Livesley, R. K., *Finite Elements: An Introduction for Engineers*, Cambridge Univerity Press (1983).

Love, A. E. H., *A Treatise on the Mathematical Theory of Elasticity*, 4th edn, Dover Publications (1944).

McClintock, F. A., and Argon A. S., *Mechanical Behaviour of Materials*, Addison-Wesley (1966).

Monro, D. M., *FORTRAN 77*, Edward Arnold (1982).

Muskhelishvili, N. I., *Some Basic Problems of the Mathematical Theory of Elasticity*, Noordhoff (1963).

Neal, B. G., *Structural Theorems and Their Applications*, Pergamon (1964).

Owen, D. R., and Zienkiewicz, O. C., Torsion of axisymmetric solids of variable diameter—including acceleration effects, *Int. J. Numer. Methods Eng.*, **8**, 195 (1974).

Rockey, K. C., *et al.*, *The Finite Element Method*, 2nd edn (1983).

Peterson, R. E., *Stress Concentration Factors*, Wiley-Interscience (1974).

Richards, T. H., *Energy Methods in Stress Analysis*, Ellis Horwood (1977).

Roark, R. J., and Young, W. C., *Formulas for Stress and Strain*, 5th edn, McGraw-Hill (1975).

Rooke D. P., and Cartwright, D. J., *Compendium of Stress Intensity Factors*, HMSO (1975).

Savin, G. N., *Stress Concentration Around Holes*, Pergamon (1961).

Sechler, E. E., *Elasticity in Engineering*, John Wiley (1952).

Southwell, R. V., *Theory of Elasticity*, Dover Publications (1969).

Spillers, W. R., *Introduction to Structures*, Ellis Horwood (1985).

Timoshenko, S. P., and Gere J. M., *Mechanics of Materials*, 2nd edn, Wadsworth (1984).

Timoshenko, S. P., and Goodier, J. N., *Theory of Elasticity*, 3rd edn, McGraw-Hill Kogakusha (1970).

Timoshenko, S. P., and Woinowsky-Krieger, S., *Theory of Plates and Shells*, 2nd edn, McGraw-Hill Kogakusha (1959).

Wilson, E. L., Structural analysis of axisymmetric solids, *AIAA J.*, **3**, 2269 (1965).

Zienkiewicz, O. C., *The Finite Element Method*, 3rd edn, McGraw-Hill (1977).

Zienkiewicz, O. C., and Cheung, Y. K., Finite elements in the solution of field problems, *Engineer*, **507**, September (1965).

Zienkiewicz, O. C., and Cheung, Y. K., Stresses in shafts, *Engineer*, **696**, November (1967).

Index